D0484177

ALSO BY DAVID A. PRICE

Love and Hate in Jamestown

THE PIXAR TOUCH

THE PIXAR TOUCH

THE MAKING OF A COMPANY

DAVID A. PRICE

ALFRED A. KNOPF NEW YORK 2008

THIS IS A BORZOI BOOK PUBLISHED BY ALFRED A. KNOPF

 PUBLISHED IN THE UNITED STATES BY ALFRED A. KNOPF, A DIVISION OF RANDOM HOUSE, INC., NEW YORK, AND IN CANADA BY RANDOM HOUSE OF CANADA LIMITED, TORONTO.

WWW.AAKNOPF.COM

KNOPF, BORZOI BOOKS, AND THE COLOPHON ARE REGISTERED TRADEMARKS OF RANDOM HOUSE, INC.

LIBRARY OF CONGRESS CATALOGING-IN-PUBLICATION DATA

PRICE, DAVID A.

THE PIXAR TOUCH : THE MAKING OF A COMPANY / BY DAVID A. PRICE.

P. CM.

INCLUDES BIBLIOGRAPHICAL REFERENCES.

ISBN 978-0-307-26575-3

1. PIXAR (FIRM) 2. ANIMATED FILMS—UNITED STATES.

3. COMPUTER ANIMATION. I. TITLE.

NC1766.U52P75 2008

384.8065'73—DC22

2008000165

MANUFACTURED IN THE UNITED STATES OF AMERICA

FIRST EDITION

Character animation isn't the fact that an object looks like a character or has a face and hands. Character animation is when an object moves like it is alive, when it moves like it is thinking and all of its movements are generated by its own thought processes. . . . It is the thinking that gives the illusion of life. It is the life that gives meaning to the expression. As {Antoine de} Saint-Exupéry wrote, "It's not the eyes, but the glance — not the lips, but the smile."

—JOHN LASSETER, 1994

All progress depends on the unreasonable man.

—GEORGE BERNARD SHAW, 1903

CONTENTS

THE PIXAR TOUCH

1. ANAHEIM

Like much of Southern California, the place where John Lasseter was standing had been part of an orange grove only fifty years before. The morning of March 10, 2006, Lasseter was on-stage at the Arrowhead Pond arena in Anaheim, the venue of the Walt Disney Company's annual shareholder meeting. In front of him was an audience of four thousand or so investors, ranging from retirees to young parents who had brought along their children.

A couple of miles to the west was Disneyland, itself built on a former orange grove. Lasseter had worked there during summer breaks in college, first as a sweeper in Tomorrowland, whisking away candy wrappers and popcorn, and later as a guide on the Jungle Cruise. About twenty miles to the north was Whittier, his childhood hometown; there, he'd started his professional art career at the age of five with a crayon drawing that won him fifteen dollars in a contest at a local market. A bit farther, around forty-five minutes up Interstate 5, was the old Disney animation building in Burbank, where he'd worked as a junior animator some twenty-two years ago.

Lasseter presented a strikingly laid-back appearance, as usual. He wore blue jeans and tennis shoes; his wireless mike peeked out from beneath the collar of a Hawaiian-style print shirt. Between his attire and his round, cherubic face, it was hard to remember that he was a forty-nine-year-old senior executive of a company that had just been bought for $7.4 billion.

The past half-dozen years, under former chief executive officer Michael D. Eisner, Disney had set its shareholder meetings in a series of cold-weather cities during cold-weather months—the 2005 meeting had been held in Minneapolis in February—all the better, many suspected, to discourage attendance by the company's increasingly restless shareholders. This year, however, there was no need for such ruses. In October 2005, chief operating officer Robert Iger had replaced the divisive Eisner as CEO. On taking the helm, Iger had quickly carried out a bold stroke, negotiating a buyout of Pixar Animation Studios. Although the price was high, the acquisition had proven to be overwhelmingly popular.

The deal was the culmination of one of the great reversals of fortune in business history. Pixar had started life twenty years earlier, not as Pixar Animation Studios, but as Pixar, Inc., a computer hardware company, one of hundreds dotting the landscape of the San Francisco Bay Area. A tiny animation group, consisting of Lasseter and a few technical guys, made short promotional films that brought in zero revenue; the group came close more than once to being shut down. When the Pixar Image Computer proved unprofitable, the company turned to selling animation software and making animated commercials for the likes of Listerine and LifeSavers. It kept running losses year after year.

In 1991, when Disney agreed to finance the production of Pixar's first feature film for theatrical release, Disney had all but dictated the terms of their contract. In late 1993, Disney ordered Pixar to shut down production of the Lasseter-directed film, *Toy Story,* on account of script problems—leaving Pixar's management and staff to wonder whether there would ever be a Pixar feature film at all. Many in Hollywood wondered whether audiences would even sit through a full-length film in the new medium of computer animation. It was unclear whether computer animation would prove to be a sterile novelty that would quickly wear thin.

Before long, however, Pixar was no longer the supplicant. A succession of beloved and commercially successful feature films, starting with the 1995 release of *Toy Story* and also including *A Bug's Life, Toy Story 2, Monsters, Inc., Finding Nemo,* and *The Incredibles,* made Pixar the world's dominant animation studio. Disney made a fortune distributing the films. When Pixar's majority owner and CEO, Steve Jobs, cut off contract talks with Disney in early 2004, as the end of their distribution contract was in sight, the business press focused on the question of how *Disney* would cope. By then, it was Disney that seemed to need Pixar—the distribution of Pixar's films made up around 45 percent of the operating income from Disney's film operations. Disney's internal market research showed that mothers of young children trusted the Pixar brand more than Disney's.

With the acquisition, Lasseter was soon to hold the newly confected titles of "chief creative officer" for Disney and Pixar animation and "principal creative advisor" for Disney's worldwide theme parks and resorts. The success of Pixar's films had brought him something exceedingly rare in Hollywood: not the house with the obligatory pool in the backyard and the Oscar statuettes on the fireplace mantel, or the country estate, or the vintage Jaguar roadster—although he had all of those things, too. It wasn't that he could afford to indulge his affinity for model railroads by acquiring a full-size 1901 steam locomotive, with plans to run it on the future site of his twenty-thousand-square-foot mansion in Sonoma Valley wine country. (Even *Walt Disney's* backyard train had been a mere one-eighth-scale replica.)

None of these was the truly important fruit of Lasseter's achievements. What success had brought him, most meaningfully, was freedom. Having created a new genre of film with his colleagues at Pixar, he had been able to make the films *he* wanted to make, and he was coming back to Disney on his own terms.

Lasseter was on hand that morning in Anaheim to whip up enthusiasm for the acquisition and for Pixar's forthcoming films, *Cars* and *Ratatouille.* Iger had teased the audience earlier in the morning with a bit of misdirection, pretending to build up to an introduction of

Lasseter and then changing the subject. When Iger was finally ready to bring him on, Lasseter met with energetic applause and whooping and yelling as the sound system played a few bars of the signature song of *Toy Story* and *Toy Story 2,* "You've Got a Friend in Me."

When Lasseter speaks onstage, his exuberant gestures and delivery make it seem as if he's filling the room. His voice, although slightly nasal, is still warm and appealing.

"For the Disney stockholders' meeting, my wife said, 'You've got to dress up, John,' " he began, "so I wore *black* tennis shoes." The crowd laughed and gave him another smattering of applause.

He recalled his high school afternoons, running home to catch Bugs Bunny cartoons on channel 11. He told of finding the book *The Art of Animation* by Bob Thomas, and realizing that some people actually earned a living from making cartoons. "That's what I wanted to be," he said.

He remembered working on the Jungle Cruise ride, and the corny bits of comedy that the ride operators tossed off to enliven the show. Elsewhere, he had told friends that the experience brought him out of his shell and gave him confidence in front of people. Apparently, it worked.

"My favorite joke is—remember the natives that are on the pole and the rhino was there? You'd come up and say, 'Oh, there they are, it's the famous Hontas tribe. They've been lost for a long time. Oh, look, the rhino is trying to . . . poke-a-Hontas!' "

He recounted joining Disney's animation division fresh out of the California Institute of the Arts, and seeing one of the early trials of 3-D computer animation. "I knew that Walt would have loved this," he said.

He elided over the disappointment he'd felt as a young man in trying, and failing, to get the studio to embrace the new technology. Keeping his presentation upbeat, he said simply, "So I followed my dream to work with one of the most amazing people I've ever known in my life—Ed Catmull."

The sixty-one-year-old president of Pixar, Dr. Edwin Catmull, looking on from the audience, was a computer graphics researcher who had fallen accidentally into the role of business leader. With his salt-and-pepper beard and round gold-rimmed glasses, he still resembled an academic. He was a geek's geek who routinely brought scientific books and papers with him on vacation as recreational reading. Those who knew him well invariably described him as brilliant, but his intelligence was not the kind that advertised itself by dominating conversations; by and large, he preferred listening to talking. When he did speak, his words were genial and courtly.

While Catmull had built upon the work of others, the revolution in 3-D animation was in many respects his revolution. It was Catmull who had held to his youthful vision of fully computer-animated movies. It was Catmull, running computer graphics labs at the New York Institute of Technology and George Lucas's Lucasfilm, who had assembled the technical and creative teams that would become the core of Pixar Animation Studios.

"In 1986, Steve Jobs bought us from Lucasfilm, and we formed Pixar," Lasseter went on. "For the first ten years, we lost a lot of Steve's money. A *lot* of Steve's money."

Steve Jobs was missing from the scene of the meeting, though he would soon be Disney's largest individual shareholder (the acquisition was a stock-for-stock trade) and the newest member of Disney's board. Lasseter was right about his money; Jobs had driven a hard bargain in buying Lucas's Computer Division for five million dollars (not ten million, as is sometimes reported), but as it turned out, he put ten times that amount into the company over the course of a decade to keep it afloat. Few other investors would have had the patience of Jobs.

Lasseter continued by relating his affection for all things Disney and showing off footage from *Cars* and *Ratatouille.* He closed by bringing his story full circle: He was proud, he said, to come home to Disney again.

Listening to Lasseter's presentation in its full sweep, one could recognize in it the cadences of a Pixar film: stretches of adventure narrative and comedy punctuated by moments of disarming earnestness—all capped by the inevitable happy ending. With the benefit of hindsight, Pixar's success story does seem inevitable. The studio enjoyed an unprecedented six box-office triumphs in a row during its years as an independent company, turning out hits with clock-like regularity, and it would continue its winning streak under Disney's wing.

Yet Pixar's story was anything but preordained. It is a triple helix of artistic, technological, and business struggles, and it is a study in the tremendously uncertain and contingent nature of artistic, technological, and business success. It illustrates how professional prestige and social status flow into each other, and how a small organization can magnify its power by deploying them as an economic force. It shows how small things, done well, can lead to big things. It is the story of a small group of individuals who started with a shared ambition to create a new way of telling stories—within a virtual world of mathematical constructions—and who traveled a long and circuitous road until their vision became a reality.

Just as Pixar's success was far from preordained, neither was the success of any of its leading figures. Indeed, one of the curious aspects of Pixar's story is that each of its leaders was, by conventional standards, a failure at the time he came onto the scene. Lasseter landed his dream job at Disney out of college—and had just been fired from it. Catmull had done well-respected work as a graduate student in computer graphics, but had been turned down for a teaching position and ended up in what he felt was a dead-end software development job. Alvy Ray Smith, the company's co-founder, had checked out of academia, got work at Xerox's famous Palo Alto Research Center, and then abruptly found himself on the street. Jobs had endured humiliation and pain as he was ejected from Apple Computer, the company he had co-founded; overnight, he had transformed from boy wonder of Silicon Valley to a roundly ridiculed has-been. For each of these figures, one could say that failure rewarded him by opening the way to new and ultimately grander outlets for his ambitions.

The odysseys of these figures, and of Pixar as a whole, bring to mind the observation of the maverick economist Joseph Schumpeter that successful innovation "is a feat not of intellect, but of will." Writing about the psychology of entrepreneurs in the early twentieth century, a time when the subject was unfashionable, he believed few individuals are prepared for "the resistances and uncertainties incident to doing what has not been done before." Those who braved the risks of failure did so out of noneconomic as well as economic motives, among them "the joy of creating, of getting things done, or simply of exercising one's energy and ingenuity." In Pixar's case, at least, the resistances and uncertainties were abundant—as was the will.

The story begins as a soft-spoken, straitlaced former missionary named Ed Catmull enters graduate school at the University of Utah in 1970. At the time, the discipline of 3-D computer animation is still an orange grove.

2. IN THE GARAGE

Now and then in history one finds a time and a place that seems to be charmed, where talent has assembled in a way that appears to defy all laws of probability—drama in Elizabethan London, philosophy in Athens during the third century BC, painting in late-fifteenth-century and early-sixteenth-century Florence. One of the lesser known among these is Salt Lake City in the 1960s and early 1970s—to be more precise, computer graphics at the University of Utah computer science department. Ed Catmull came of age professionally in this milieu, and his career would for decades to come show Utah's influence on both his ambitions and his managerial style.

The university had recruited Dave Evans, a computer science professor at Berkeley and an elder of the Church of Jesus Christ of Latter-day Saints, to start its computer science department in 1965. Evans, in turn, recruited other top researchers to join the faculty—most notably, a thirty-year-old tenured Harvard professor named Ivan Sutherland. For his doctoral thesis at the Massachusetts Institute of Technology, Sutherland had developed a system called Sketchpad, pathbreaking for its time, that allowed users to sketch black-and-white engineering drawings with a light pen and a computer display. It wasn't just that Sutherland's system let a person draw with a computer, which was mind-blowing enough. At a time when users stood in line to drop punch-card programs into a card reader, the better to meter out the computer's precious milliseconds, Sketchpad relied on the *insane* idea of letting one person use a room-size computer *all by himself.*

Equipment for computer graphics could not be bought off the shelf in the late 1960s, so Evans and Sutherland also started a company to build and sell it. Consequently, researchers at Utah always had the latest top-of-the-line graphics hardware. Money flooded into the department from ARPA, the Advanced Research Projects Agency, a unit of the Pentagon created in response to Sputnik to fund research on next-generation technologies.

Computer graphics was a small field, and the word spread that Utah was the place to go. Evans, uninterested in the hierarchy and formalities of graduate education, wanted students undertaking their own research as soon as possible; the more daring the objective, the better. He and his colleagues gave graduate students autonomy and respected them as professionals, treating them as top-flight researchers who just happened not to have Ph.D.'s yet.

The result was an environment in which it was not unusual for students to achieve major advances. Henri Gouraud, a young Frenchman, came to Utah's department and developed a superior way to calculate the shading of curved three-dimensional objects, one that made the objects look far smoother than had been possible before. The Vietnamese-born Bui Tuong Phong came to Utah and was the first to conceive a method for creating objects with realistic-looking illumination and highlights. (Gouraud shading and Phong lighting are still used in graphics software today.)

Other doctoral students of the period would go on to play key roles in the development of computer graphics and in laying the foundations of modern personal computers. The list of names reads like a Homeric roll call: Alan Kay (Ph.D., 1969) would conceive, among other things, object-oriented programming and the point-and-click graphical user interface, both of them now ubiquitous. He would advocate in the early 1970s for the bizarre (as it seemed at the time) concept of a notebook-size computer that you could carry around with you. John Warnock (Ph.D., 1969) would do pioneering work with digital typefaces and desktop publishing, leading to his co-founding Adobe Systems. Jim Clark (Ph.D., 1974) wrote his thesis on virtual-reality displays, founded Silicon Graphics to build computers for high-speed 3-D graphics, and co-founded Netscape at the dawn of the era of the World Wide Web. Nolan Bushnell, an undergraduate (B.S., 1969), would go on to start Atari and popularize video games. Another Dave Evans recruit on the faculty, Tom Stockham, would pioneer both digital enhancement of photographs and digital sound recording.

Catmull entered this Athens more or less by circumstance. He was a local boy out of Granite High School across town. Born on March 31, 1945, in Parkersburg, West Virginia, he had grown up in Salt Lake City, the oldest of five children in a Mormon family. He had come to the university as an undergraduate in 1963, interrupting his college years to go on the two-year missionary experience

The Granite High School debate squad, 1962–1963. Ed Catmull, captain, is at right. In addition to debate, Catmull was active in the math, science, and drama clubs. As a boy, he had hoped to become a Disney animator, but he decided in high school that he lacked the ability to draw.

Granite High School yearbook photo

that was customary for young Mormon men—the church sent him to Coney Island and Scarsdale, New York.

The interruption proved to be providential indeed: He returned to campus just as the computer science program was getting underway. Alan Kay, then a graduate student, taught Catmull's first programming course. Kay remembered Catmull for his habit of going beyond the call of an assignment for fun. "He just liked to program, and usually would add things to the assignments that he thought about along the way," Kay said. "This is always a good sign."

After graduating in 1969 with a degree in computer science and physics, Catmull took a job at Boeing, but was soon caught up in a mass layoff along with thousands of other employees. He came back to the department as a doctoral student with a radical idea. He had aspired in his youth to be a Disney animator; as a boy, he looked up to Walt Disney and made flip books in preparation for what he assumed would be his future career in cartoons. In high school, however, he had ruefully concluded that he couldn't draw. Now the thought seized him: Computers might allow him to do animation after all. With computer graphics, he could create not just individual images, but feature-length animated films.

"Computer animation was sort of on the lunatic fringe at that time," said Fred Parke, a fellow Ph.D. student in Catmull's class who also worked on animation. "People were just barely to the point where they could get a computer to put out still images."

It was obvious that it would take years for the state of the art in computer hardware to catch up with this ambition, and there was no end of problems to be solved on the mathematical and programming side. Nonetheless, from Catmull's point of view, there was no better time or place to get started than right here and now.

In 1972, Catmull took the opportunity to make a short animated clip for a graduate course project. He decided to digitize and animate the closest thing at hand—his left hand. Nothing about the film came easily or simply. He began by making a plaster-of-paris mold of his hand; when he pulled the mold away, the hair on the back of the hand came out painfully with it. He then made a plaster model from the mold and drew about 350 small triangles and other polygons on the model in ink.

When he was done, the drawings of the polygons crisscrossed the plaster hand like a net. Just as one could approximate a curved line with a series of short straight lines, a curved object could be represented by a mesh of polygons. Digital counterparts of these polygons would represent the surface of his hand in the computer. He laboriously measured the coordinates of each of the corner points of the polygons and typed them into the computer with a Teletype keyboard. With a 3-D animation program that he wrote, he could reproduce the disembodied hand on a screen and make it move.

Just getting a look at his imagery was a task in itself. Because the display hardware never showed the entire image on screen at any one moment—it took thirty seconds or so to cycle through the image—Catmull could see a frame of his work only by taking a long-exposure Polaroid of the screen and looking at the snapshot. Once he was satisfied, he then shot the footage using a thirty-five-millimeter movie camera that the department had rigged to take pictures from a CRT screen.

The resulting film, roughly a minute long, was jaw-dropping in its day. It showed the hand swiveling, opening, and closing, and then—the pièce de résistance—the film took the viewer up and into the inside of the hand and looked around. Parke created a computer animation of his wife's face at around the same time. The two films defined the cutting edge of computer animation for years afterward. (Snippets of both films were also incorporated into the now little-remembered 1976 film *Futureworld.*)

Professor Sutherland opened up a line of communication with the Walt Disney Co. to see whether Disney could be persuaded to use computer graphics in its production process for traditional animation. Knowing of Catmull's enthusiasm for animation, Sutherland brought Catmull to Disney to meet with executives there. Disney management wasn't interested in computer graphics, but invited Catmull to help the Disney Imagineering team use computers to design a new ride—specifically, Space Mountain, a roller-coaster ride planned for the Walt Disney World complex that had just opened in Orlando, Florida. Catmull begged off and the two men went back to Salt Lake City.

Catmull's thesis project had him tackling three-dimensional curves once again. He was interested in bicubic patches, a mathematical way to represent curved surfaces that was smoother than a polygon mesh with all its edges. He figured out how a computer could decide which parts of an object created from bicubic patches should be visible to the viewer, and which parts were hidden from view (because they were behind something else). It was one of those problems in computer graphics that sound easy in principle, but turn out to be very hard in practice. Catmull's solution to the hidden-surface problem relied on an invention he called the Z-buffer, an area of memory that kept track of the distance between the viewer and the closest surface at each point of a scene.

His work on bicubic patches and the Z-buffer was groundbreaking in itself. Yet his thesis had still another major discovery in store. As an offshoot of his fiddling with the mathematics of curved objects, he found a way to project an image, any image, onto an object's exterior. With this invention, known as texture mapping, an object in a computer graphic could be given an appearance of marble, say, or a wood-grain finish. His first texture map was a projection of Mickey Mouse onto an undulating surface. In his dissertation, he illustrated the concept with a projection of Winnie the Pooh and Tigger.

The trifecta of accomplishments in Catmull's 1974 doctoral thesis—bicubic patches, the Z-buffer, and texture mapping—would have ensured him lasting fame in the field of computer graphics even if he had never done anything else. In his mind, though, they were just stepping-stones to his real goal—namely, computer-animated feature films.

The path forward for a newly graduated computer graphics specialist was not well marked. Neither academia nor corporate America—nor, of course, the entertainment industry—was hiring battalions of people with Catmull's expertise. Catmull was turned down for a faculty position at Ohio State University, which had a small computer animation effort underway. He then put his hopes on a new project of Professor Sutherland, who had just gone to Hollywood to start a second firm, a production company for computer animation called the Picture/Design Group. Its goal was to create computer-animated footage on assignment to be incorporated into commercials, television shows, and films. Catmull waited in Utah while Sutherland and his partners sought investors.

By this time, Catmull had a wife, Laraine, and a two-year-old son depending on him. Months went by without a job opening up for him at Sutherland's company, and finally the Picture/Design Group shut down altogether. Practicality prevailed, and Catmull accepted a programming job in Boston with a company called Applicon. The company made software for computer-aided design, an area far from Catmull's interests. The graphics work at Applicon was based on line drawing, not the solid 3-D geometry and shading that Catmull cared about. Character animation, of course, was well off the company's agenda. At the age of twenty-nine, he was distressed to find himself simply marking time—doing what he perceived to be dull, routine work, yet unsure what direction to go.

This had continued for several months when he received a puzzling phone call at the office. On the line was a secretary for someone he had never heard of. She was calling to set up Catmull's airline ticket to New York, she told him.

"I had no idea who the woman was or why she was trying to get me to New York," Catmull recalled.

Unknown to Catmull, an eccentric multimillionaire had appeared on the scene back in Salt Lake City. Alexander Schure, a successful entrepreneur, had founded a university called the New York Institute of Technology in 1955. Schure was its president. Lately, Schure had been bitten by the moviemaking bug and had opened an animation studio on one of the school's campuses on Long Island. Watching the animators at work, he was struck by the amount of costly hand labor involved, and came to believe that computers could replace most of the people.

Catmull's animated hand, plus the animated face of Mrs. Fred Parke, was now part of a promotional reel for Dave Evans and Ivan Sutherland's hardware company, Evans & Sutherland. An enterprising salesman for E&S had been making cold calls on East Coast universities and had struck gold when he gave a copy of the film to Schure. Soon Schure was in Utah meeting with Professor Evans and Professor Sutherland and buying one of every kind of equipment they had to offer.

In the course of the shopping extravaganza, Evans asked Schure who was going to run his computer graphics operation. "Who should it be?" Schure asked in return.

"Well, you just missed the right guy," Evans told him. "Ed Catmull has just taken another job out of desperation."

Once Catmull understood why Schure's secretary was calling, his interest perked up. He flew to New York and found that Schure was offering him exactly the break he had been wishing for: his own research lab devoted to computer animation. During the interview, Catmull asked Schure what he'd bought from Evans & Sutherland. Schure replied that he didn't know—all he knew was that he had one of everything.

In November 1974, Catmull became director of the NYIT Computer Graphics Lab. He brought along his officemate from Applicon, Malcolm Blanchard, who started in January.

The school eschewed academic aspirations of the ivy-covered variety. Somewhere between a third-tier university and a diploma mill at the time, it prospered by meeting the needs of two groups of students: G.I. Bill veterans who could not gain admission elsewhere, and young men who needed a student exemption from the draft to keep them out of the Vietnam War.

Catmull's computer graphics lab had no involvement to speak of with the rest of the university. The setting on Long Island's North Shore was as peaceful as it was photogenic. It was F. Scott Fitzgerald country. Schure had assembled the campus from several grand estates and used the mansions as the school's buildings. For the lab, Schure set aside a converted four-car garage—albeit an unusually stately garage—in the shadow of one of the mansions. Catmull began setting up a workroom in the garage's upper level, formerly the chauffeurs' quarters, and a computer room on the ground floor.

His next task was to gather a team to push the limits of 3-D graphics. The group he brought together at NYIT would evolve into Pixar Animation Studios, which was, in a hyperliteral sense, a garage company.

At the same time, another aspiring computer artist was undergoing his own odyssey some three thousand miles away. Alvy Ray Smith, born in the panhandle of Texas and raised in New Mexico, had a lifelong love of painting. He had also picked up computer programming while he was a college student at New Mexico State University; the school's computer classes were taught by men from nearby White Sands Missile Range.

Like Catmull, he didn't feel he could support himself with his art, so he went to Stanford on scholarship to study for his Ph.D. in electrical engineering. He continued to paint in his spare time, exhibiting at the Stanford Coffee House at one point. He finished his doctorate in 1968, having also absorbed the politics of the Bay Area of the late 1960s. He went on to New York University to teach a branch of computer science called cellular automata, the field in which he had done his thesis work—the mathematics of self-reproducing machines.

Smith might have continued along his sure and steady academic progression if not for a momentary slip of his ski cap. He was schussing down a slope in New Hampshire in 1973 when his headgear fell over his eyes and blocked his view of another skier heading toward him, wildly out of control. He spent the next three months stuck in a full-body cast, chest to toes.

"This time turned out to be one of the most wonderful of my life," he remembered.

It forced him to rethink everything. He realized that his life of comfort and relative privilege was on the wrong track, "not using my artistic talent, not enjoying the fact that only a dozen people in the world could talk CA [cellular automata] with me."

And he realized something else that troubled him. "I was supporting the war effort, which I disapproved of, by teaching computer science," he said. "Those guys went off and helped the military machine."

He decided to go back to California with no job and no plan. He would live on a shoestring, relying on his savings until he happened upon the right situation. "I just knew, somehow, that something good was going to happen if I could get back out to California."

Smith settled in Berkeley and waited for whatever was going to happen.

After some months, he took on a writing project that required him to spend time at Stanford's library, forty miles away. He cadged a room for the night from a friend, Dick Shoup, who was living in Palo Alto. Over lunch the following day, Shoup invited Smith to come to his office to see the software that he had written for painting. Shoup was based at Xerox's Palo Alto Research Center, or PARC—a collection of Ph.D. researchers that Xerox mostly left alone to work on audacious technology projects for the office of the future. PARC's farsighted work under Alan Kay on personal computers and graphical interfaces had already made it the first corporate research lab, in all likelihood, to be profiled in *Rolling Stone* magazine. Other teams at PARC were working on office computer networks and developing the first laser printer.

For the past several years, knowing that Smith was an artist, Shoup had been trying to interest Smith in his painting-software project. Smith, however, could never make sense of Shoup's notion of painting with a computer—it seemed like gibberish—and had always put him off. Now, having been Shoup's houseguest, he could not decently avoid the invitation again.

What Smith had awaiting him was a look at the first color painting program anywhere. Shoup called it SuperPaint; it used a tablet and stylus and, despite its pioneering status, it offered many of the garden-variety functions of modern painting software. As Shoup put the system through its paces, Smith was dumbstruck. He recorded the visit in his journal: "I was reluctant to go. . . . I went because of his hospitality—and was greatly and very pleasantly surprised! His machine finally exists: a color TV 'paintbrush' hooked up to a computer. It is dazzling."

Smith left the lab realizing he had found the "something good" he was waiting for. He came back a few days later and, losing all track of time, ended up staying on the machine for twelve hours. He implored Shoup to get him on board at PARC, somehow, so he could keep experimenting with computer painting. Shoup couldn't win approval to hire him, but was able to circumvent the company's systems by paying Smith through a purchase order—as if Smith were a case of staplers.

Smith started in August 1974. His job was to create a video of animation that would show off the system's capabilities. He mixed abstract sequences with classic walk animation that he picked up from a how-to-animate book. Another young artist, David DiFrancesco, found his way to PARC, and the two took turns running wild on SuperPaint.

Xerox took away their plaything in January 1975, however, following a corporate decision to concentrate on black and white. Smith told anyone at Xerox who would listen: Hey, look, color is the future and you own it completely. But the decision was final—color had no place in the office of the future. Smith's purchase order was canceled.

The main thing Smith had lost was not the paycheck, or even the painting program itself (he could write one if he had to), but the graphics hardware that was under the hood. It was a device called a frame buffer, a region of computer memory that one could, in effect, look at on a screen. The frame buffer was the canvas of Shoup's painting system. Each point, or pixel, on the computer's video display corresponded to a memory location in the frame buffer. When a program changed the number that was stored in some memory location of the frame buffer, the color or shade of gray of the corresponding point on the screen instantly changed with it. A paint program such as SuperPaint essentially followed what the artist did with the stylus and then changed the frame buffer's memory accordingly. Unlike earlier graphics hardware, which could only draw lines, a frame buffer gave the programmer complete control over what appeared on screen.

Thus, to paint with a computer again, Smith and DiFrancesco would have to find another frame buffer. This was no small task. Although omnipresent in computers today, they were hardly anywhere to be found in 1975; Xerox PARC had had to design and build its own. Rumor had it that the University of Utah was getting one, though, so the two took off in Smith's white Ford Torino for Salt Lake City.

When they reached Utah's computer science department, Smith and DiFrancesco assiduously avoided using the word *art,* correctly guessing that the department and its Pentagon-funded projects had no need for artists. Nonetheless, the people there quickly sorted out what the two were after. There was nothing at Utah for them. But a graduate student mentioned that there was another place they could try: An oddball multimillionaire from Long Island had just come through and ordered everything in sight—even an eighty-thousand-dollar frame buffer.

With their hey-man style and down-to-there hair, Smith and DiFrancesco freely marked themselves as Northern California hippies. As Smith showed his excitement about what he had just heard, he received a friendly warning. This guy from Long Island hired one of our Ph.D.'s to run his lab, Smith was told: Ed Catmull. He's good, but he's a real strong Mormon—very straight. Watch out.

Undeterred, Smith and DiFrancesco spent the last of their money on plane fare to New York. There, they borrowed an old Porsche from DiFrancesco's father and made their way through a snowstorm to the converted garage that housed Catmull's little operation.

Catmull welcomed them and explained his assignment: to do whatever needed to be done to make animated films on a computer. Elsewhere on the campus, Schure had hired a team of more than a hundred animators, background artists, and so forth from Hollywood and New York to work on a traditional, hand-drawn animated film; Catmull's group, as soon as he had one, was to learn animation from them.

Outwardly, Catmull and Smith were a study in contrasts: Catmull, a slender man and reserved; Smith, outgoing and outspoken, built big in a way that reminded some of Smokey Bear. Still, they got along easily, sharing each other's excitement for computer graphics and its seemingly boundless possibilities. Catmull, as it turned out, wasn't put off at all by Smith's Bohemian nature; the warning in Utah had been needless. "He was just accepting," Smith recalled. "He didn't lay his trip on anybody. And he didn't discourage you from your trip."

Catmull, feeling overwhelmed by his project, was pleased to have an offer of help from the visitors. The four of them—Catmull and his former officemate Blanchard, the 3-D guys, plus Smith and DiFrancesco, the 2-D guys—were soon in a limousine carrying them to a mansion on the next estate over. They were ushered across the mansion's cavernous foyer and entered the dining room. As uniformed waiters scurried about, the men heard a voice from a table across the room. "Welcome, California!" roared Alexander Schure.

Schure was a true visionary of computer animation, staking a fortune on the idea at a time when the concept was difficult for others even to fathom. Part and parcel of his visionary nature, as Catmull already knew and Smith was about to learn, was his unorthodox manner of communication. It was a fluid, impressive-sounding rhetoric that often fell slightly short of making sense. DiFrancesco would later call it a "word salad." (A notorious specimen of Schure's verbal style was his pronouncement to a reporter, "Our vision will speed up time, eventually deleting it.")

"He would start talking and we didn't know what he was talking about," Smith remembered. "He just didn't speak in the usual conversational exchange. He would spew this poetry and, in order to get a word in, you had to start talking—just start talking at the same

The former garage and chauffeurs' quarters on Long Island's North Shore where Pixar had its origins as the Computer Graphics Laboratory of the New York Institute of Technology.
Courtesy of Alvy Ray Smith

time. And after a while, if you heard your words coming back in his flow, you figured, well, I guess that idea transferred."

The frame buffer had not shipped yet from Evans & Sutherland, so the men started by teaching themselves how to use the 3-D line-drawing system that had already come in, as well as the lab's minicomputer, a PDP-11/45 from Digital Equipment Corp. A friend told them about a newfangled operating system called Unix that had just come out of AT&T's Bell Labs, together with the programming language that came with it, called C. Catmull and Smith both disliked Fortran, the programming language then pushed by IBM. Fortran was a symbol, in their eyes, of bland big-company mediocrity; they quickly became devotees of what they regarded as C's more logical and elegant style.

Once the men had the technical basics in place, the NYIT Computer Graphics Lab was an ocean of opportunity and freedom: Your job was whatever you thought was important, so long as you were filling in a piece of the computer animation puzzle. Catmull worked on a 2-D program called TWEEN to automate the so-called inbetweening of traditional hand-drawn animation—that is, to draw frames of character motion between an animator's hand-drawn frames. (For a while before Unix came in, Catmull wrote it in assembly language, the computer's most tedious, most laborious, lowest-level language—such was his aversion to Fortran.) After the frame buffer arrived, Smith worked on a program for painting backgrounds, taking notes from a background painter in Schure's cel animation group. DiFrancesco experimented to find better methods for filming images from a screen. Blanchard, a master systems programmer, fixed bugs in Unix and the C compiler.

Catmull gradually brought more people on board. Employee number five, Christine Barton, one of the first women in the field, went to work on creating a network among the lab's computers. (The network had to be built from scratch; this was years before local networks would reach the mainstream.) Jim Clark continued the work he had started at Utah on virtual reality; his device was a headset display, basically two small screens mounted near the user's eyes to project computer graphics. Another guy was tackling real-time digitizing of video—turning a video signal into a stream of binary computer data as the camera rolled (or the videotape unspooled).

A summer intern at the lab from Utah's doctoral program, Jim Blinn, worked on an effect he called "bump mapping." Catmull's texture mapping, despite its name, did not give physical texture to the surface of an object; rather, it was like painting or wrapping the surface of the object with a two-dimensional image. One could, for instance, use texture mapping to wrap an image of concrete onto a 3-D object to create the appearance of concrete, but the surface would still lack the up-and-down dimensionality of concrete. Blinn was seeking to overcome this limitation by finding a way to apply a three-dimensional texture to an object's surface, making it rough, embossed, ridged, or whatever was needed.

As the staff grew, the managerial side of Catmull's job grew along with it. His style was to try to re-create the atmosphere of an academic department—Utah's—and so the result was a loose-knit collection of largely self-directed projects. His role, as he practiced it, was to empower others: to offer counsel when asked, to run interference with the university, to handle issues with Schure. Dictates from the top were next to nonexistent. The style was a natural fit for the lab's talented and self-motivated staff, with their ardor for the lab's work.

Indeed, for most of the group, day and night did not exist there. Catmull, a family man, kept a regular daytime schedule, but the others generally worked for as long as their bodies could carry them. They were acutely aware of the privileged existence they were leading—the privilege of tapping into their shared passion for computer graphics—and wanted not to miss one more hour of it than they had to. They were intense, if not maniacal.

Music was a must. Whoever was in charge of the music for the day would put on some Pink Floyd or Cream, perhaps, or Bob Dylan, or occasionally soft jazz. Now and then, someone would say, "Oh, wow," and others would hurry over to see what the person had just accomplished. At around three or four a.m., they would catch some sleep back in their rooms—Schure had arranged living quarters for them on other nearby estates—then wander into the office in groups to start again. Smith found that he had a twenty-six-hour cycle, his workdays drifting in and out of phase with Catmull's over the course of a couple of weeks.

"It had the great sense of being this fraternity of geeks," recalled Ralph Guggenheim, a hire out of Carnegie Mellon University.

Adding to the fantasyland aura of the place was Schure's munificence in buying equipment. From a computer aficionado's point of view, it was hot-rod heaven. When Digital Equipment Corp. came out with a next-generation minicomputer called the VAX, the NYIT graphics lab got the first machine off the assembly line—notwithstanding the price tag of more than $200,000 (over $600,000 in present-day dollars). "He kept asking us what we should get next; we'd tell him and he'd buy it," Smith said.

A crucial moment in Schure's buying spree came after the

researchers told him it would be nice to have two more frame buffers. Talking over and through his word salad, they said that adding two frame buffers to the one that the lab already owned would allow much better images; the three devices could be tied together, giving triple the amount of memory per pixel. A few weeks later, during a routine visit to the upstairs workroom, Schure casually mentioned, "Oh, I just bought you five more frame buffers."

It was crazy-wonderful. Schure had dropped three hundred grand (in mid-1970s dollars) on the basis of an offhand suggestion. The machines he had bought would give the lab an unprecedented capability. The key to it was a simple relationship: The more memory one had per pixel, the more colors or shades of gray one could choose from at any point on an image. If a frame buffer offered one bit, one binary digit, per pixel, the choices for each pixel were just 0 and 1—on or off. With two bits per pixel, one could choose from any of four numbers—00, 01, 10, or 11, meaning 0, 1, 2, or 3—and thus any of four possible colors or shades of gray.

An Evans & Sutherland frame buffer offered eight bits of storage per pixel, enough so that each pixel could be set to any of 2^8, or 256, colors (or any of 256 shades of gray). A mere 256 colors was not enough for any sort of realism, however.

With three of the devices combined, though, one frame buffer could handle 256 levels of red; another, 256 levels of green; and the third, 256 levels of blue. The images could thus have up to 256 × 256 × 256 colors, or more than 16 million—a complete color spectrum, as a practical matter. This was the exotic power, commonplace today in the cheapest of digital cameras, that Schure had suddenly bestowed on them. The lab was now the first on the planet to be able to pursue photorealistic computer imagery. (At any rate, it was the first civilian lab with that ability; no one knew what the military and intelligence agencies might be doing behind their closed doors.)

There were other benefits. Lines or edges drawn on a computer often take on a jagged appearance—a staircase effect where the line is supposed to be smooth. The imperfections in the lines were and are commonly known as jaggies (or, more formally, as aliasing). The effect of jaggies was even worse in animation, where they created a crawling-ants effect along edges. One way to deal with them was to mix the color of a line and the colors of the areas next to the line in many combinations to create an illusion of smoothness. It wasn't feasible with 256 colors, which did not allow enough color choices for effective blending, but a sixteen-million-color palette meant you could make lines look the way they were intended to look. Catmull believed that banishing jaggies was vital to creating graphics that would be acceptable to movie audiences. Finally, the fact that Schure had bought five of the devices meant the lab would have six of them altogether—enough for *two* glorious full-color displays.

The access to the latest equipment was energizing in itself, but the long hours were mainly in the service of a vision, namely, that they could one day make movies. "From the very get-go, they were working toward the point where they could be Disney," said Ted Baehr, a friend of several of the staffers and a frequent visitor to the lab. "That's all they talked about."

Even when the group broke away from work for the occasional evening drive into Manhattan, the goal of moviemaking was rarely far out of mind. When the Walt Disney Co. screened a series of its classic animated films over the summer of 1976, the NYIT computer graphics team devotedly took them in. Other times, they might attend a lecture at the New School for Social Research, where they heard critic Leonard Maltin hold forth on film appreciation and the history of animation. Every summer, they would take a road trip to a computer graphics conference called SIGGRAPH, where they presented technical papers and their own brief animated clips.

Their relationship with the idea of animation was better than their relationship with the actual animators at NYIT. Schure had his hand-drawn animation group at work on a film based on the children's story *Tubby the Tuba.* Although computer graphics researchers were able to glean much about the mechanics of cel animation techniques from them, the *Tubby the Tuba* contingent became wary as time went on. Schure had a habit of scaring them with loose talk about his concept of computers eliminating animators—no matter how often Catmull and Smith tried to explain to him that it wouldn't work that way, that computer animation would still require animators. "Alex would say to the artists, 'Someday, you guys aren't going to have a job because these guys are going to replace you,' " Catmull remembered, "and we knew that wasn't true."

Tubby would turn out to be a career counselor of sorts. In the spring of 1977, with the film completed, Schure hosted a private showing in a screening room at the Manhattan offices of Metro-Goldwyn-Mayer. The film had a veteran cast, including Dick Van Dyke as the voice of the eponymous tuba. The director, however, was a first-timer—Alexander Schure himself.

Catmull and his group found it painful to watch. Anything that could have gone wrong, it seemed, did go wrong. There was dust on the frames, there were shadows under the lines, the music was annoying, the story grated, and the entire production was simply boring. Smith, in the front row, could take no more and closed his eyes. A programmer on the computer graphics team fell asleep. After the lights came up, a distraught young animator exclaimed, "I've just wasted two years of my life!"

For the computer fanatics, it was a moment of revelation. For all their lofty aspirations, Ed Catmull and Alvy Ray Smith had been focused only on the technological side of filmmaking. "I don't think anybody [in the computer lab] was thinking in terms of plot and story and telling stories," Baehr observed. "In the beginning, it was just, 'We want to make a movie.' "

The calamity of *Tubby the Tuba* forced them to confront an unpleasant fact—namely, that they were in the wrong place for making good movies. Money was not enough, they could now see. Technical genius was not enough (though *Tubby* had grave technical problems, too). Splendid equipment would not be enough. For them to make worthwhile films someday—not just the R&D exercises they showed at SIGGRAPH meetings—there also had to be people on board who understood film storytelling. Schure, although blessed with great foresight, could not be their Walt Disney.

It was a lesson that went back to the early days of animation. A technically gifted animator named Ub Iwerks had helped Walt Disney Productions (as it was then called) establish its leading position in the 1920s, and he had even been primarily responsible for the design of Mickey Mouse. Iwerks was a master of fitting motion to music, of innovative camera moves. But when he left to start his own studio, his work consisted of mediocre cartoons starring now-forgotten characters like Flip the Frog and Willie Whopper. Without Walt Disney's grasp of character and drama, Iwerks's mechanical talent was of little value. So, too, the wizardry of the NYIT computer graphics team could not, on its own, yield compelling cinema.

What to do and where to go next, however, were vexing questions. Catmull and Smith had already been making annual pilgrimages to Disney to try to get the company interested in their group. Disney was the only animation studio with the resources to support a group like theirs; besides, the men were Disney fans. One subterfuge or another was always necessary to keep Schure in the dark about the visits—Smith would nominally be traveling to Florida, say, and Catmull would be on his way to San Francisco, yet somehow, they would both end up in Burbank. They made the rounds of other major studios, too, but they assumed Disney would be the one that would bite if anyone did.

As in Catmull's graduate school days, however, the Walt Disney Co. was not interested in computer graphics. Walt had died of cancer in 1966, and the company was now run by a caretaker chief executive, Esmond Cardon "Card" Walker. Some of Disney's technology experts saw great promise in the NYIT group's work, but that was as far as it ever went.

Who else had pockets deep enough to support a major research effort into computer animation for filmmaking? It might take a decade, or even longer, before computer costs came down enough for a feature film to be anywhere near the realm of possibility. The only option, it seemed, was to keep making progress on the technical issues—on NYIT's dime—while waiting for Disney to call.

A call did come in, but not from the place they expected.

In early 1979, Ralph Guggenheim was part of a four-person group within the lab assigned to work on television commercials; Schure had come to believe the lab should start to carry some of its own weight with paying projects. The group had created a visual effect for a Chevrolet ad the previous year and had done a test for Royal Crown Cola (they didn't get the job). A commercial for a local stereo store was in production. Guggenheim was relaxing in his apartment one evening after work when the phone rang. The caller identified himself as head of development for George Lucas. His name was Bob Gindy.

Lucas had sent the man on a quest—namely, to find someone who understood the mysterious force of computers and who knew how to apply that force to filmmaking. The quest had been difficult. Gindy's path had taken him from one computer science professor to another, with no one able to help him—until he found a computer scientist at Carnegie Mellon named Raj Reddy, who told him that Guggenheim had done work on computer animation when he was in school.

Gindy explained to Guggenheim that Lucas wanted to modernize the tools of filmmaking. From Lucas's point of view, filmmaking technology seemed to be frozen in time. Film editing still meant cutting and splicing film stock manually. Preparing sound tracks meant weeks of cutting and splicing magnetic tape and combining sounds on a manual mixing board; sometimes you needed two or three guys to control the mixing board's faders.

Combining images for special-effects work, positioning everything accurately frame after frame, was the worst of all. For the lightsaber scenes in *Star Wars,* animators had to create images of glowing lightsabers with meticulous airbrush work so the movement of the weapons would follow the wooden dowels held by the actors; the images then had to be incorporated into the film through a process known as optical compositing. Some of the shots of spacecraft required special-effects crews to composite forty or fifty layers of film. Getting one shot right could take months. Lucas wondered whether computers might be the answer.

Help us, young Ralph. You're our only hope.

"At the same time, he was telling me what a great place Marin County [Lucasfilm's home] is, how beautiful it is, how great the real estate values are, that sort of thing," Guggenheim recalled.

Guggenheim asked him what his area was at Lucasfilm. "Are you a computer guy?"

"Oh, no no no," Gindy replied. "I'm head of development here—real estate development. I buy property for George. I don't know anything about computers; none of us at Lucasfilm really knows anything about computers. That's why we want to put this group together."

Outsiders assumed that Lucasfilm must be bursting with computers. In fact, Lucas's special-effects team had used computers only to control the motion of the spaceship models. An outside computer graphics artist named Larry Cuba had been hired to create the footage of the Death Star attack plan that Luke Skywalker watched in the briefing room. All the other imagery in *Star Wars* that appeared to be computer-generated, such as the computer readouts and the Death Star countdown, was actually created on analog video equipment or with traditional animation.

Gindy said he wanted to know whether Guggenheim might be interested in running a new computer research lab at Lucasfilm. The lab would tackle the three projects Lucas was interested in—film editing, sound, and compositing. Also, the company needed a good accounting system. A year and a half after the release of *Star Wars,* Lucas thought it might be time for his company to start using computers for accounting.

"Based on what Raj Reddy told me, you would be the guy," he said.

Few places held more cachet in 1979 than Lucasfilm. To be the company's head of computer research would be a dream position. Nonetheless, Guggenheim made a snap decision to count himself out. Less than a year out of a master's program, he didn't have the experience. Ed Catmull, he realized, was the one Lucas really needed.

"Wow, I'm really honored," Guggenheim told Gindy, "but I'm working with the geniuses of the field here—people who have years more experience than I do. Let me talk very confidentially to a few people. I think I know who you need."

That would be fine, Gindy said. He just had one question. "What we need to know more than anything else is this. I've talked to a bunch of people who are doing this sort of thing and none of them can answer this *one question* to my satisfaction."

What's the question? Guggenheim asked.

"Do you guys have the ability to take a spaceship and make it fly around on the screen?"

Guggenheim felt relieved. "We do that every day," he replied. As it happened, they had just been doing a test of something similar that same week.

The next morning, he started to tell Catmull and Smith about a call from George Lucas. "Close the door!" Smith interjected.

As Guggenheim related the conversation, Catmull was flabbergasted. Everyone in the lab was in awe of *Star Wars.* One weekend in the summer of 1977, during one of their excursions to Manhattan, the group took in a matinee of the film; it amazed them so much, they saw it again that same day. They assumed they could only dream of getting a call from Lucasfilm.

"I don't know if you're aware of this," Catmull said, "but Alvy and I have been taking time every summer to go to L.A. and visit all the major movie studios and try to talk them into doing just this."

They would have to move with extreme caution. When Schure had learned earlier that Jim Clark was looking for a job elsewhere, Schure fired him on the spot. Just how Schure had found out in the first place remained a mystery—he even brandished copies of Clark's e-mails. Catmull figured someone within the lab must have hacked Clark's e-mail and informed on him, but it wasn't clear who.

So Catmull kept a lid on the news. Catmull and Smith, loath to take a chance on another e-mail breach, rented a manual typewriter and spent hours drafting and redrafting a letter to George Lucas.

Shortly afterward, Bob Gindy flew out to NYIT together with Richard Edlund of Lucas's special-effects operation, Industrial Light & Magic. They were there incognito; apart from Catmull, Smith, Guggenheim, and one or two others, no one at the lab (now up to thirty or so) was told who they were or where they were from. The small group had prepped a demo for the visitors—a fancier version of the animated spaceship. Catmull and Smith next made two furtive trips to California to meet with Lucasfilm's president, Charlie Weber, and other Lucasfilm executives. The second time, Lucas squeezed in ten minutes or so with Catmull while waiting for the setup of a special-effects shot on *The Empire Strikes Back.*

Catmull passed Lucas's test. He accepted the job of head of Lucasfilm's newly created Computer Division. But he couldn't bring his group over en masse; Lucas was creating a place only for him, at first. Catmull would have to draw up plans and make the case for hiring more people once he was there.

Before Catmull left, he and the small in-the-know group at NYIT spent hours figuring out strategy—how to avoid panic within the group, how to mitigate any negative reaction from Schure against those who would still be working there. As Catmull and Guggenheim were regular racquetball players, they took to discussing the move on the racquetball court, where they enjoyed the privacy that was lacking in the lab. It became a sort of code: When it was time to talk about Lucasfilm, one of them would suggest heading out for a racquetball game.

What emerged was a plan to keep Schure from suspecting what was afoot, and possibly suing. The departures from NYIT to Lucasfilm would take place gradually and indirectly. Those who were interested in going to Lucasfilm would take interim jobs elsewhere—"laundering" themselves, as the group called it; Catmull would bring them in when he could.

The plan went without a hitch. Catmull gave notice and started at

Lucasfilm in July. Smith and DiFrancesco went to the Jet Propulsion Laboratory (JPL) in Pasadena for several months to work with Jim Blinn on graphics for Carl Sagan's mini-series *Cosmos.* Guggenheim stayed at NYIT until May 1980, keeping a one-year commitment he had made when Schure hired him, then went to a company in Pittsburgh for a few months. All told, a half-dozen people from Catmull's team at NYIT made their way over during the course of a little more than a year.

Schure was puzzled and hurt by the exodus. "He had a sense that he had been betrayed, that they had sort of taken advantage of him," recalled Fred Parke, Catmull's classmate, who became director of the lab several years later.

Catmull, for his part, had given NYIT five years of his career and had made the lab a world center of graphics research. Now he had a new patron—one who was, at the moment, perhaps the most famous filmmaker on the planet. True, Lucas and his executives had never said anything about the Computer Division actually creating films, but that detail could be dealt with later.

3. LUCASFILM

In the fall of 1979, Ed Catmull had an office in a small two-story building that Lucas owned in San Anselmo, a quaint town in Marin County. The first floor, rented out, was an antiques store; Catmull shared the upstairs with the director's wife and film editor, Marcia. That the two of them ended up in adjoining spaces was happenstance, but it was a fitting combination: Marcia Lucas was supervising the design work on a grand filmmaking center to be built on a rural 1,882-acre spread using *Star Wars* money. Skywalker Ranch, as it was called, was one of Lucasfilm's ambitious, expensive, long-term projects to remake film production. Catmull's Computer Division was the other.

Catmull's command of the mathematics and technology of 3-D computer graphics had given him a margin of comfort at NYIT. At Lucasfilm, he started with no such advantage. The truth was that he did not have a terribly deep grasp of the areas Lucas wanted him to work on. No one alive, probably, had a handle on all of them. The technologies that Lucas was looking for—digital film compositing, digital audio mixing and editing, and digital film editing—existed, for the most part, only in Lucas's own fertile imagination. Catmull would just have to dive in and cope.

Alvy Ray Smith, mastermind of the 2-D painting programs at NYIT, was the logical choice to lead the digital compositing project when he and David DiFrancesco finished laundering themselves at JPL in early 1980. Smith's group, which initially consisted of himself and DiFrancesco, faced a seemingly endless horizon of question marks: how to build a machine that could scan movie film at high resolution, with exacting precision from frame to frame; how to combine "bluescreen" images in a computer to produce natural-looking results; and, perhaps hardest of all, how to get images from a computer onto movie film with cinematic quality. No one even knew how much resolution was needed for the digitized imagery to look good to an audience. Taking photos of a CRT screen, the tried-and-true method for getting computer images onto film, might or might not be good enough.

On the brighter side, Catmull was witnessing the power of the Lucasfilm cachet. Not for the last time in his career, he found that the chance to work for a high-status organization—high-status in the sense of *cool*—attracted top people as much as money did, if not more so. Anyone whom he approached was eager to come on board. When he and Smith determined that the top person in the world at the time for digital audio was a Stanford professor named Andy Moorer, they drove to Palo Alto to see whether they could talk him into leaving the university to run the sound editing project. As they entered his office and introduced themselves, he got out of his chair and said, simply, "If you guys are here for what I think you are, the answer is 'yes.' "

Others importuned them. Résumés poured into Catmull's and Smith's offices. A young animator named Brad Bird, recently graduated from the California Institute of the Arts, hung around trying to get a computer animation project going.

"He was one of the funniest guys on Earth," Smith said. "We'd just hang out, talk, dream. He had all these ideas for making animated movies, but he didn't have a technical bone in his body and he didn't have any tolerance that you would need to have at the time to put up with some of the awfulness of the early technology."

Bird would be back two decades later.

In Seattle, a programmer named Loren Carpenter, a thirteen-year veteran at Boeing, heard about the new computer operation at Lucasfilm and pondered how to get in. Just tossing his résumé in over the transom would not do it. "They could heat their building by burning all the résumés they were getting," he said.

Carpenter had long wanted to work in computer graphics and had already made his way into Boeing's computer-aided design group, which created software for designing and rendering airplanes and airplane parts. "It was the closest thing that Boeing had," he recalled.

Carpenter set about making himself stand out for Lucasfilm. He persuaded his manager at Boeing to let him use the company's computers on nights and weekends to create a short animated film, one that he could present at SIGGRAPH in August 1980, four months away. The film would use fractals, a branch of mathematics that the French-born mathematician Benoît B. Mandelbrot had just lifted out of obscurity. Carpenter realized that fractals, which reproduced many of the patterns found in nature, would let him achieve a new level of realism in depicting the ridges and crags of mountains. The film was thus to take the viewer on an airborne passage over and around a fractal mountain range, as if from the perspective of a small plane. He called it *Vol Libre*—free flight.

As the film ran at SIGGRAPH, the audience of engineers and academics was awestruck. Catmull and Smith, sitting in the front row, offered him a job on the spot.

By the fall, Catmull's team had outgrown its space over the antiques shop; the group moved to a converted laundromat a block away. The projects were still in the planning phase as the researchers drafted white papers and design documents. Oddly enough, the Lucasfilm Computer Division did not yet have a computer, or even a word processing machine. The only typewriter was on the desk of Catmull's secretary (and future wife), Susan Anderson; anyone else who wanted to compose on a keyboard had to wait for her to go to lunch. It was hard to envision that this was the group that would spearhead the computer revolution in filmmaking.

When Lucas came around to the office, there was an implicit rule: No gawking at the boss. "He wanted to feel like he's one of the guys," Ralph Guggenheim recalled. "He didn't want people to genuflect to him or make a big deal out of his presence."

Catmull had all the leaders for Lucas's projects in place by this time. Smith was in charge of the compositing project; under him, DiFrancesco tinkered with laser technology, trying to figure out how to build a digital film scanner and digital film printer. Andy Moorer was in charge of sound. Guggenheim had been brought into the fold to run the editing project. Catmull hired someone to work on the accounting system; he arranged for the person to report to Lucasfilm's financial managers to get that project off his own plate.

That wasn't the end of the agenda, however. As Catmull and Smith saw it, there were the goals Lucas had set down, and then there was the one he *ought* to have, but didn't—namely, computer animation. To Catmull and Smith's frustration, although Lucas now had on his payroll perhaps the world's top technical talent in 3-D animation, he wasn't asking the group to do any. His special-effects group at Industrial Light & Magic saw no use for computer graphics, either. Indeed, from the standpoint of computer graphics, Lucas's second *Star Wars* film, *The Empire Strikes Back* (1980), was a step backward from the first: It included no computer graphics at all. "It became pretty clear fairly early on," Smith remembered, "that there wasn't a lot of enthusiasm that we thought there would be inside Lucasfilm for computer graphics."

The breakthrough came in the fall of 1981 when Paramount Pictures contracted with Industrial Light & Magic to work on *Star Trek II: The Wrath of Khan.* In an effort to snag ILM's attention, Catmull's group had created a short film in August, a scene of the starship *Enterprise* chasing a Klingon battle cruiser.

It worked. Jim Veilleux of ILM, visual-effects supervisor on the film, asked for a meeting. Veilleux invited Smith to help with a key sequence in which Admiral Kirk (as he now was), Mr. Spock, and Dr. McCoy would view a simulation of the "Genesis device"—an experimental unit meant to be shot into the surface of a dead moon or planet and bring it to verdant life.

Smith expressed enthusiasm, but gave Veilleux a caveat. "We can't do movie resolution yet," he explained. "We can only do video resolution."

That wasn't a problem, Veilleux said. The effect was going to appear on a video monitor on the *Enterprise.* Video resolution would be fine.

Veilleux suggested that the sequence could show an aquarium-like tank with a rock floating inside; the Genesis device would cause plant life to form on the rock. This, Smith felt, was aiming far too low.

"Do you guys know what you can do with computers and what you can't do?" he asked. It was a rhetorical question. "I know the parameters of your shot now," he went on. "Let me think about it overnight and I'll come back to you."

When they parted, Smith was in a state of elation; he knew it was the big break. "I basically went out of there dancing," he recalled.

That night, he thought about the planetary flyby shots he had worked on with Jim Blinn at JPL. He thought about the way George Lucas seemed to watch a film; he had noticed that Lucas, unlike a mortal moviegoer, paid close attention to camera moves and all the decisions that the director made about the camera. Smith stayed up all night drawing and redrawing storyboards.

He emerged with the concept of a scene shown from the perspective of an imaginary spacecraft as it approached a dead planet and flew past. The audience would see a sperm-shaped missile hitting the planet with the Genesis device, which would spark some sort of violent, chaotic process sweeping across the surface. With Carpenter's fractal techniques, the scene would show volcanoes exploding and mountains forming. The camera would then pull away to show the post-Genesis planet, now resembling Earth.

After some back and forth, Smith was able to sell ILM and Paramount on the plan. He brought his team together and told them they had the contract—they were in the movie business. The project, he told the team, would be "a sixty-second commercial to George Lucas." The scene would fit Paramount's purposes; the razzle-dazzle would not be gratuitous—it would make narrative sense. But their real mission in making the scene was to telegraph to Lucas what they could do for him.

There was much to be done by the scene's due date of March 19, 1982. Carpenter animated the fractal mountains and mapped the camera's complex, twisting flight path. Several team members—Tom Duff and Tom Porter, hired in late 1980 and early 1981, and newcomer Rob Cook—wrote software to generate cratered textures for the surface of the dead planet and to allow an ILM matte artist to paint the planet in its final, Earth-like form. By this time, the graphics team was in larger quarters and had two VAX computers, which were soon working around the clock on computations for the scene.

Carpenter lobbied successfully for an added touch of verisimilitude. It would not do, he felt, just to make up stars. He suggested that the planet should be in orbit around the real-life star Epsilon Indi, which some thought to be a likely location of life-bearing planets; it is around 11.3 light-years distant from Earth. Using data from the Yale Bright Star Catalogue, a reference with information on the positions and magnitudes of ninety-one hundred stars, he designed the star field in the scene to correspond to the actual stars that would be visible from Epsilon Indi.

The nature of the chaos on the planet proved hard to work out. Bill Reeves, a Ph.D. from the University of Toronto who had been working on another project, took everyone by surprise by unveiling a revolutionary rendering technique that he'd created in his spare time. His technique, called "particle systems," dealt with clouds of thousands of particles; it made it possible for the first time to create realistic animations of phenomena like fire, smoke, water, and dust. Each particle was like an organism with rules determining when it was born, how it moved and changed form during its lifetime, and when it died. The particles could move at random to produce a more natural-looking effect.

To demonstrate, Reeves created a few animations of fires. Like Carpenter's mountains, they looked more striking and more realistic than anyone had seen up to then. Smith and Carpenter, trying to brainstorm a solution to the chaos problem, agreed that it made sense to replace the exploding volcanoes with a ring of particle-system fire that would race across the planet.

The production schedule was punishing, and it became more so as the group had to move offices in midstream, from an industrial park in Novato to a Lucasfilm complex on Kerner Boulevard in San Rafael. Nonetheless, the "sixty-second commercial" wrapped on time. The day after the first in-house screening of *Star Trek II,* George Lucas appeared at the door of Smith's office, said "Great camera shot!" and was gone as suddenly as he had come.

When *Star Trek II* released on June 4, the Genesis sequence wowed audiences. It was not the first use of computer animation in a feature film; there had already been computer-animated footage in *Futureworld* (including Catmull's and Fred Parke's student projects) and Larry Cuba's work in *Star Wars,* as well as computer-animated displays in *Alien* (1979) and a computer-animated simulation of the actress Susan Dey in *Looker* (1981). Of all the computer animation in feature films up to that point, however, it was arguably the most dramatic visually. Excerpts of it reappeared in the next several *Star Trek* films; it was the effect that would not die.

Embedded in the Genesis sequence—on screen for anyone to see, yet visible only to the expert eye—was Catmull's philosophy of computer animation. From a technical point of view, Catmull believed fervently that computer animation had to meet the expectations set by people's perceptions of everyday life. This did not mean it had to achieve photographic realism; it did, however, have to avoid certain spell-breaking effects that would trigger the audience's rejection—even though the effects might be so subtle that the viewer could not articulate what was wrong.

One of these was movement without motion blur. In old stop-motion effects, which were made up of images of still figures, the absence of motion blur gave the footage a strobelike or staccato feel that made it look false. ILM had solved the motion-blur problem for *Star Wars* by putting the spacecraft models under the control of computers, which caused them to move as footage was shot; the result was that motion blur appeared on film.

Computer animation was susceptible to the same problem as stop-motion; Catmull believed it was essential for the group to counter the problem by finding algorithms for adding realistic motion blur to their films. Audiences, he believed, would never accept computer animation without it. In 1982, the group still hadn't found a general solution to the motion-blur problem, but it had been able to incorporate some blurring here and there in the Genesis sequence (in the star field and Reeves's fires).

The other phenomenon that Catmull had in mind was jaggies, the staircaselike imperfections that his group had been exorcising from its images since NYIT. Jaggies had well-known mathematical solutions, which were known generically as anti-aliasing. The problem was that people in computer graphics sometimes didn't bother with anti-aliasing, or would treat it as an afterthought. At Lucasfilm, Catmull insisted that all software written by the group had to deal with jaggies from the outset; anyone who failed to comply would, he decreed, have their "ballzen shieren offen."

Catmull's main competition at the time in the feature-film arena, the team of John Whitney, Jr., and Gary Demos, took a different tack. To the casual observer, the two teams looked about evenly matched. Whitney and Demos had an equally weighty set of film credits: They had created the ersatz Susan Dey for *Looker* as well as a computer model of Peter Fonda's head for *Futureworld;* they also contributed some of the computer imagery to the science-fiction film *Tron,* released a month after *Star Trek II.* Like Catmull and Smith, they aspired to create computer-animated feature films when the technology was ready and the costs were within reach. On the question of jaggies, however, the two groups embraced opposite approaches. Whitney and Demos believed the answer was higher resolution, not fancy anti-aliasing algorithms. First at a company called Information International, Inc. (or Triple-I, as it was usually known), then at their own firm, Digital Productions, their approach was to create imagery with thousands of lines of resolution using high-end hardware. With more detailed images, the theory went, the jaggies would become smaller and the problem would just go away.

It was a clever theory—clever, plausible, and too good to be true. Jaggies turned out to be objectionable even at high resolution. Catmull had been right: You couldn't just throw more expensive hardware at the problem. You had to work out the fundamentals.

It so happened that ILM had talked with Whitney and Demos in 1978 before approaching the NYIT group; Richard Edlund of ILM later showed Catmull and Smith the X-wing fighter images that Whitney and Demos had generated to try to win their way into the Lucasfilm empire. They were very high-resolution images, executed beautifully—but they didn't sell ILM. The damn things had jaggies.

While work continued on Lucas's several computer projects, Catmull and Smith looked for new opportunities for the graphics group to take on production work. Despite the success of the group's scene in *Star Trek II,* the only motion-picture job for the time being was the next *Star Wars* sequel, *Return of the Jedi* (1983). For a briefing scene, Bill Reeves and Tom Duff created a mock hologram of the moon of Endor (home of the teddy-bear-like Ewoks) and a half-completed Death Star nearby.

The group also collaborated on an exercise meant to test the limits of their ability to create lifelike imagery. Completed in April 1983, the image was titled *The Road to Point Reyes,* a play on the name of a nearby state park and their rendering program. (Reyes, the name of the program, doubled as an acronym for "Renders Everything You Ever Saw.") In the foreground was a two-lane country road, its pavement texture-mapped by Rob Cook; on the side of the road were grasses created with Reeves's particle systems and flowering plants designed by Smith; beyond were a lake and mountains, created with Carpenter's fractals. Others contributed as well. The group referred to the image as a "one-frame movie"—a sardonic reference to the long list of credits, but also, implicitly, an acknowledgment of the dearth of *real* movies for the group to get its hands on.

A twenty-six-year-old Disney animator approached Catmull around the same time to find out whether the group could create a 3-D forest. The animator was working on a proposal for a film of Thomas Disch's story "The Brave Little Toaster," which had just appeared in *The Magazine of Fantasy & Science Fiction.* The idea was to combine traditional, 2-D animated characters with some computer-animated sets. The animator met with Catmull and Smith at Lucasfilm to discuss his project; they returned the visit, spending time with him at Disney Animation in Burbank. There, after hours, he took Smith to the archive in the basement and thrilled him by pulling out original drawings from the drunken-elephant scenes in *Dumbo* and Preston Blair's hippo–alligator dance in *Fantasia.*

Catmull and Smith liked the fellow and were impressed by his interest in computer graphics—a rarity in professional animators, they had found. Nothing ever came of the project, however. (Disney would release a conventionally animated version of *The Brave Little Toaster* in 1987.)

Lacking any film production assignments, Catmull and Smith decided to take matters into their own hands. In July, they returned from the annual SIGGRAPH conference determined to make a splash at the following year's show—lots of papers, a prototype of Guggenheim's film editing system, and a short film of their own creation. Officially, the latter project was intended to test the group's new rendering algorithms, including what they hoped was a solution to the motion-blur problem. Unofficially, the film would give the group the production work it wanted and, ideally, persuade George Lucas that they could create good cinema.

Fare at SIGGRAPH film showings was heavily weighted toward reels of digital effects from TV commercials, spinning-logo network identifications (the portfolios of production houses seeking to drum up business), and demonstrations of researchers' new rendering algorithms. An emerging community of computer artists rounded out the schedule with avant-garde art films that tended to strike technical types in the audience as convoluted and irritating. Smith wanted Lucasfilm's piece to contrast with all this by centering, somehow, on a character. He drew storyboards with a stick-figure android character named André waking up in a forest. For the title, the team members settled on *My Breakfast with André,* a twist on the Louis Malle film that many of them loved.

In November, Catmull gave a talk at a computer graphics conference on the *Queen Mary.* (The retired ocean liner was permanently docked at Long Beach.) He encountered the Disney animator at the conference and asked how things were going with *The Brave Little Toaster.*

"It got shelved," John Lasseter said glumly.

Catmull asked what he was up to. Lasseter told him he had been laid off from the studio between projects. Catmull understood. At ILM, many of the staff had "run-of-show" employment; when a production was done, they were off the payroll until the next film got underway—which was usually right around the corner. It was no big deal.

Lasseter could not bring himself to tell Catmull the truth: He would not be working on the next production at Disney, or the one after that. He had been fired.

Catmull phoned Smith later that day to go over business. As the conversation wound down, Catmull mentioned, "I ran into John Lasseter. He's not working at Disney."

Smith urged him to put down the phone and hire Lasseter *right now.* The next thing Lasseter knew, he heard a voice from behind a column in the back of the ballroom. It was Catmull's. "John, John," Catmull said in a stage whisper. "C'mere. *C'mere . . .*"

John Lasseter was born in Hollywood, California, on January 12, 1957, to Paul and Jewell Lasseter and was raised in Whittier. He grew up at a time when cartoons were understood to be only for children, and adolescents were expected to shed their interest in such juvenilia. He never did. "I even watched them when it wasn't cool in high school," he recalled later. "I quietly ran home after school to watch cartoons—*Bugs and His Buddies* on KTTV channel 11."

He attended Whittier High School with twin sister Johanna and older brother Jim. During his freshman year, he was browsing in the school library when he found Bob Thomas's *The Art of Animation.* The book gave him a glimpse of the workaday world behind Disney animation—the background painters, the layout men, the story men with their storyboards, the women of the Ink and Paint Department, the directors, and, above all, the animators. "The key man in the art of animation has been the animator," the book declared. "He always will be."

The book opened Lasseter's eyes to a new idea: People had jobs creating animation.

Soon afterward, he went to a local theater (now the Whittier Village Cinemas) to see a forty-nine-cent showing of Disney's *The Sword*

John Lasseter in his senior year of high school. By this time, he had already been determined for several years to work for Disney. His senior class voted him "best artist."
Whittier High School yearbook photo

in the Stone. He was wary of the social suicide that would result if any of his classmates witnessed him going to a Disney cartoon, so he had his mother drop him off at the theater alone. When she picked him up afterward, he had an announcement for her: He wanted to work for Disney.

His mother, an art teacher at another high school, told him it was a fine goal.

He started sending letters and drawings to the studio and getting notes of encouragement back. In his senior year, Lasseter received a letter from the California Institute of the Arts (or CalArts, as it is commonly called) inviting him to apply to its new character animation program. The opportunity might as well have been tailor-made for him. He spent the summer of 1975 as an assistant to the program's director, Jack Hannah, helping with photocopying and the like, then started school in the fall.

CalArts was Walt Disney's brainchild; he had started the planning of the school in the late 1950s and provided generously for it in his will. Walt and his brother Roy formed it in 1961 through a merger of two struggling Los Angeles institutions, the Los Angeles Conservatory of Music and the Chouinard Art Institute. The doors opened at the school's consolidated campus in Valencia in 1971, five years after Walt's death. (The school's first board chairman, H. R. "Bob" Haldeman, had been the front-runner for the presidency of Walt Disney Productions before he left in 1969 for a star-crossed tour of duty in the Nixon White House.)

The school had nearly been finished off before Lasseter got there. The Valencia campus opened at a time when hippiedom was near its peak. The Disney family, although forward thinking in many ways, was uncomfortable with the protest movements and flower-child culture that inevitably seeped into the school. The final straw came during a meeting of the school's campus affairs committee, when a member of the photography faculty showed up in the nude to protest an edict against skinny-dipping in the swimming pool. Once word of the incident got out, the family and the board decided enough was enough. The board's then-chairman, Harrison Price, was dispatched to make an offer to Justin Dart, his counterpart at the University of Southern California: USC could have the school, lock, stock, and barrel, plus eighteen million dollars in endowment money. Dart, a hard negotiator by reflex, demanded twenty-four million. The Disney family, now more infuriated by Dart than by the hippies, brought the talks to an end and recommitted itself to the school.

The storms of the 1960s had mostly receded by the time Lasseter arrived. At CalArts, he found his own kind of liberation: Here, he no longer needed to conceal his passion for cartoons. His twenty classmates from across the country were animation geeks like him. Others had been corresponding with the Disney studio just as he had, and even making their own short films. Many would go on from CalArts to perform significant work at Disney or elsewhere; among them were future stars John Musker (co-director of *Aladdin, Hercules,* and *The Little Mermaid*) and Brad Bird.

First-year classes took place in room A113, a windowless space with white walls, floor, and ceiling, and buzzing fluorescent lights. The teachers made up for the setting, however: Almost all of them were longtime Disney artists with awe-inspiring animation credits. Kendall O'Connor, an art director on *Snow White and the Seven Dwarfs,* taught layout; Elmer Plummer, a character designer on *Dumbo,* taught life drawing; T. Hee, a sequence director on *Pinocchio,* taught caricature. The program was rigorous and the hours long; the fact that the campus was in the middle of nowhere made it easier to focus on work. Tim Burton, who entered the program the following year, remembered the experience:

> It was like being in the Army; I've never been in the Army, but the Disney program is probably about as close as I'll ever get. You're taught by Disney people, you're taught the Disney philosophy. It was kind of a funny atmosphere, but it was the first time I had been with a group of people with similar interests.

The school's library had sixteen-millimeter prints of a half-dozen Disney features. Groups of students would watch the films repeat-

John Lasseter (back row, with pencil in his mouth) and his entering class at the California Institute of the Arts character animation program, March 1976, the spring semester of the first year. His classmates, back row (L–R): Joe Lanzisero, Darrell Van Citters, Brett Thompson, Leslie Margolin, Mike Cedeno, Paul Nowak, Nancy Beiman. On platform: Jerry Rees, Bruce M. Morris, Elmer Plummer (Life Drawing instructor, with beard), Brad Bird, Doug Lefler. Front: Harry Sabin, John Musker. Not pictured: Tim Barker. Tim Burton was in the following year's class.
Courtesy of Harry Sabin

edly, stopping them or slowing them down to study interesting moments, looking to understand how the masters worked.

"We wore those prints out analyzing them," Lasseter remembered. "There would be about ten of us. We would go out and have dinner and then spend the evening analyzing the films."

The class had a high rate of attrition, both outward and upward. Nancy Beiman, one of the three women in Lasseter's class, recalls that a third of the class left by 1978, either to transfer to the more free-form "experimental" animation program or to quit the school altogether. Another third got jobs at Disney before graduating. If the studio had a need and a student showed special promise, the studio would pluck him or her from the program right away. Burton was one of these, hired after his third year on the strength of a short film called *Stalk of the Celery Monster;* Musker was another. Lasseter was offered a job at Disney in his junior year, but opted to stay for his degree, earning his bachelor of fine arts in 1979. Two of his student films, *Lady and the Lamp* and *Nitemare,* won back-to-back Student Academy Awards for animation in 1979 and 1980, respectively.

Lasseter then became a junior animator at Disney, realizing his dream. The studio was of two minds about how to make use of him, however. Publicly, he was made to seem like a rising star. Soon after he arrived, the studio sent him on a circuit of campus and television appearances to project a more youthful image for the company; his articulate and approachable manner made him a natural. Yet the senior animators were skeptical of what they saw as the overeager ambitions of Lasseter and the other new CalArts hires. Even an Oscar winner such as Lasseter would require years of apprenticeship, they believed. If he wanted to make his mark in Disney animation, he would have to wait his turn.

"John's got an instinctive feel for character and movement and shows every indication of blossoming here," veteran Mel Shaw told the *Los Angeles Times* during Lasseter's first year on the staff. Shaw said that Lasseter would make a superb contribution—"in time," he noted pointedly.

"I sense these young kids are a lot more impatient to get ahead and be successful than we were at their age," said Wolfgang "Woolie" Reitherman, director of *The Jungle Book,* who had joined Lasseter on his publicity tour. "They want it all right now, and although I can't blame them, they have to learn to be patient. You can't just zoom out of school and become a director in a few short years."

Even in the studio's best times, Lasseter would likely have encountered a bumpy adjustment. The late 1970s and early 1980s, however, were not Disney's best times. What Lasseter soon found was that Disney animation—and the company as a whole—had entered a kind of dormancy. Walt Disney's portrait was ubiquitous in the hallways, in offices, in lobbies, but his genius was acutely missing. Walt Disney, the man, had a restless zeal for storytelling and a gift for harnessing the latest technology to the storytelling craft. He had been first with talking animated films; he had been first with full-color animation. He made the first American animated feature, *Snow White,* staking more than $1.4 million on it (around $19 million in present-day dollars) in the midst of the Great Depression. He had even reconceived the lowly amusement-park business as a storytelling medium.

His company was now run at its top levels by men who understood themselves to be following his example, without understanding that his example consisted of bold strokes. Thirteen years after Walt Disney's death, the most favored expression of executives seemed to be, "As Walt used to say . . ."

The company's own market research announced the problem to anyone who cared to listen. The research was spurred by declining ratings for *The Wonderful World of Disney,* once a top-rated TV show, which now consisted largely of reruns and material from the vault. The research department did some surveys and came back with bleak findings: The program's loss of popularity appeared to be just a warning of more to come; the Disney magic, as consumers saw it, was dwindling across the board.

The animation building at Mickey Avenue and Dopey Drive on the Disney lot might have seemed an unlikely place for turmoil. Many of the younger artists, however, were deeply frustrated with what they saw as the studio's cheapness and artistic timidity. A faction had rallied around one of their number, Don Bluth, whom they regarded as the studio's best hope for rejuvenation. Shortly before Lasseter's arrival, Bluth resigned and a dozen staffers followed him out the door. Ron Miller, the head of the studio—later promoted to the presidency of the company—called a staff meeting and opened with the words, "Now that the cancer has been excised . . ."

The real disease, the company's sleeping sickness, continued to take its toll. Lasseter felt stifled; the films he worked on—*The Fox and the Hound* and the short *Mickey's Christmas Carol*—seemed minor and inconsequential to him. Still, he soldiered on. "It was like my heart was ripped out," he recalled. "This was not what I always dreamed Disney was."

Lasseter left Disney to work for the London-based Richard Williams Animation Studio. (Williams's best-known works at the time were the animated titles and credits for two of the *Pink Panther* films and his Oscar-winning version of *A Christmas Carol.*) Lasseter found he was no happier there and returned to Disney within a year.

In 1981, while he was working on *Mickey's Christmas Carol,* his friends Jerry Rees and Bill Kroyer invited him to look at some early footage from *Tron.* Rees, a classmate of Lasseter's from CalArts, and Kroyer were supervising the computer animation in the film, a live-action Disney production that was to take place mostly in a computer game. It would include roughly fifteen minutes of computer imagery, far more than any feature film up to that time. The film's needs were so demanding that four different computer graphics production houses had to be hired to work on it.

In a trailer on the Disney lot, Lasseter huddled with Rees and Kroyer to look at the first computer-generated scene to come in—a race among drivers in virtual motorcycles known as light cycles. The scene had no character animation and its graphics were rudimentary, but it brought Lasseter an epiphany. The dimensionality of the scene was something he had never witnessed before. If this technology could be melded with Disney animation, he thought, he would have the makings of a revolution. Until then, three-dimensional effects in animation had required difficult, costly sessions with the multistory "multiplane" camera, practical for only a few key sequences in a film, if that. The computers could even move the audience's point of view around a scene like a Steadicam. The possibilities seemed infinite.

"I couldn't believe what I was seeing," he said later. "Walt Disney, all his career, all his life, was striving to get more dimension in his animation . . . and I was standing there, looking at it, going, 'This is what Walt was waiting for.' "

He was not able to interest the animation executives in it; they did not care to hear about new technology unless it made animation faster or cheaper. The gregarious Lasseter moved easily among people at the studio, however, and had gotten to know Tom Wilhite, the head of live-action production at Disney. "John was an operator—in a good way," Wilhite remembered.

Lasseter had earlier introduced Wilhite to Tim Burton, who was designing characters for *The Black Cauldron.* Burton's un-Disney-esque designs kept getting rejected, but Wilhite, a rare risk taker among Disney executives, took a liking to his work and gave him a sixty-thousand-dollar budget to make a stop-motion short called *Vincent.* Now, with all doors closed to Lasseter's ideas in the animation studio, Wilhite agreed to fund Lasseter's proposal for a thirty-second test film. The result combined computer-generated backgrounds (executed by MAGI, the Elmsford, New York, firm that had created the light-cycle scene) with hand-drawn character animation by Glen Keane. Based on Maurice Sendak's *Where the Wild Things Are,* the film centered on an intricate tracking shot in which the viewer followed a small dog running through a house as a boy chased him.

Wilhite also supported Lasseter in taking a further step, acquiring an option on *The Brave Little Toaster* and putting together a pitch to his superiors in animation. Lasseter's gambit of going around his own chain of command proved to be too clever, however. Ron Miller quickly rejected his idea. Minutes after the pitch meeting broke up, Lasseter got a call from a manager who told him to pack his bags. "Since it's not going to be made, your project at Disney is now complete," he was told. "Your position at Disney is terminated, and your employment with Disney is now ended."

Lasseter spent a week at Lucasfilm in December 1983, then moved in for good the following month. From Catmull and Smith's point of view, he was a fantastic hire: a real animator who believed in what

The computer graphics group of the Lucasfilm Computer Division, 1984 (L–R): Loren Carpenter, Bill Reeves, Ed Catmull (vice president, Computer Division), Alvy Ray Smith (director of computer graphics research), Rob Cook, John Lasseter, Eben Ostby, David Salesin, Craig Good, and Sam Leffler. Not pictured: David DiFrancesco, Tom Duff, Tom Porter.
Courtesy of Alvy Ray Smith

they were trying to do. "A lot of animators were frightened to death of the computers, but he got it," Smith recalled. "He could see it."

Lasseter brought with him a grounding in Disney's decades of accumulated knowledge about animating characters. Moreover, he was an ideal fit for the computer animation of the era: The state of the art in computer graphics circa 1984 was best suited for rendering inanimate objects, with their geometric shapes and simple surfaces. Lasseter had an affinity for bringing inanimate objects to life and giving them personalities of their own. *Lady and the Lamp* featured a lamp shop in which the lamps were alive and eager to be bought; in Lasseter's hands, the lamps showed believable emotions of surprise, frustration, anxiety, and curiosity. *The Brave Little Toaster* was to have been a story told from the viewpoint of sentient household appliances.

There was, however, one hitch to get past—namely, George Lucas. Doubtful that Lucas would approve of his technical team hiring an animator, Catmull and Smith obscured Lasseter's role from Lucasfilm executives by giving him the title "Interface Designer."

Lasseter's real job was to work on *My Breakfast with André.* He initially approached the film as a combination of computer scenery and hand-drawn characters, much like the *Where the Wild Things Are* test. Catmull, however, pressed him to use computer animation for the characters as well. Using Smith's storyboards as a starting point, Lasseter moved away from the concept of an android and made André a boylike figure. He added a second character, an oversized, irritable bee, and the project became *The Adventures of André and Wally B.* (Wally the bee was named for Wallace Shawn of *My Dinner with André.*)

The story was simple: André would wake up in a forest, find Wally hovering in front of him, outwit Wally and run away, then get stung as the film's finale. Lasseter, perhaps still bitter over his treatment at Disney, planned to end the film on an uncharacteristically nasty note: His first set of storyboards called for a close-up of Wally's stinger pronging André's rear end. "He had it going into the butt, and getting all damaged," Smith remembered.

The shot didn't exactly seem like SIGGRAPH material. Smith, otherwise enthusiastic about Lasseter's work, asked him to move the sting off screen. "It was just anger," Smith said. "Who knows what was driving John at the time."

With the story in final form, work began on several fronts. It was the start of a pattern that would continue for decades, with the animator pushing the technology to serve the film's creative needs, and the technology inspiring the animator's art. Smith's first 3-D model of André relied on a sphere for his head and a cone for his torso—simple geometric shapes being the only ones built in to the group's modeling software. Lasseter asked instead for a rounded shape he called a "tear drop," which could move and bend fluidly. Catmull added it.

For André's eyelids and mouth, several team members collaborated on a shape called a "bound," a slice of the surface of a sphere.

Lasseter learned the group's animation program, called MD (for "Motion Doctor"), and suggested ways to make it work more in tune with the methods of a professional animator. A change could be as simple as how the program counted things: Programmers were in the habit of counting from zero, so MD counted from zero. Animators count from one, Lasseter said. Tom Duff, who had written MD, was delighted to have a former Disney animator banging on his program and telling him how to make it better.

Lasseter was learning: If he reached out to the technologists, the technologists would reach back.

Bill Reeves wrote a program to generate the film's trees, flowers, and grasses. Loren Carpenter and Rob Cook worked on improvements to Reyes, the rendering program. Reyes would take the 3-D character models and Lasseter's animation of those models, plus all the background elements and the camera movements, and create the actual images that would make up the film.

As the work progressed, one of the team members—possibly Cook, possibly Reeves—voiced concern about Lasseter's color choices. Lasseter wanted purple leaves on the trees. Leaves aren't purple, the staffer objected. Indeed, the group had been studying magazine photos of trees for reference and they seemed to back up this observation.

Rather than simply asserting his artistic authority, Lasseter took the group to a San Francisco museum exhibit of paintings by the illustrator Maxfield Parrish, noted for his rich use of light in natural scenes. After a while, the dissenter volunteered that Lasseter was right: Leaves *could* be purple. It all depended on the light. Lasseter had showed the group that there was more to realism than was dreamt of in their technical papers.

Although the film was to be a little less than two minutes long, finding enough computer time to render the frames was a major logistical problem. The team used all five of Lucasfilm's VAX computers, three of them at night and two around the clock. Reeves finagled time on ten more VAX computers at MIT. Shortly before the SIGGRAPH conference, Catmull and Smith realized that even this was not going to be enough. Rescue came in the form of Cray Research, a supercomputer manufacturer that sold some of the most powerful computers on Earth to nuclear research labs and defense agencies. Cray hoped to sell Catmull a machine and offered a test drive of two Cray X-MPs at its computer center in Mendota Heights, Minnesota.

Their rivals John Whitney, Jr., and Gary Demos, now at Digital Productions, had already taken delivery of one of the ten-million-dollar systems, which they used to create special effects for the 1984 releases *The Last Starfighter* and *2010*. Catmull and Smith disliked the idea. They had done back-of-the-envelope calculations and decided that the move didn't make economic sense: The burden of paying for the Cray would be enormous. You wouldn't own the Cray; it would own you.

A free trial, however, was another thing. Catmull and Smith feigned interest in the machine and gladly accepted the invitation.

The 1984 SIGGRAPH was to be held in late July at the Minneapolis Convention Center. Carpenter and Cook spent the final weeks beforehand at Cray scrambling to get the last of the frames ready. As it was, a couple of shots couldn't be rendered in time, so the version readied for SIGGRAPH substituted Lasseter's hand-animated pencil drawings of André and Wally for six seconds of computer animation.

The group was excited to learn that George Lucas, who had paid no attention to the project up to that point, would be flying to Minneapolis for the premiere of *André and Wally B.* It was surprising and profoundly encouraging that he would take the time; he was, by now, in charge of a kingdom that included not only film production, special effects, and sound, but also a video game division and a phenomenally successful licensing operation. On top of that, he was supervising completion of Skywalker Ranch. Catmull and Smith later realized that his presence had less to do with any newfound interest in computer animation and more to do with the coincidence that Linda Ronstadt, his then-girlfriend, had a concert in the city around the same time. (George and Marcia Lucas had separated the previous year after Marcia took up with a stained-glass artisan working at Skywalker Ranch.)

The director, shy among strangers, did not want anyone to know he was at the screening. Lucas, Ronstadt, Catmull, and Smith entered the building in a limousine through an underground garage. Conference staff led them to an elevator and signaled when it was dark in the makeshift theater. Only then did they join the thousands already in the audience.

When SIGGRAPH audiences saw something in the theater program that they liked, the applause and the hollering could get intense. From the long opening shot in which the camera trucked through a gorgeous three-dimensional forest of 46,254 Parrish-inspired trees, to the closing shot of Wally's mangled stinger, the SIGGRAPH crowd loved *André and Wally B.* and went wild when the closing credits came up.

An executive with a computer graphics company came up to Lasseter afterward and told him that the film was amazing. "What software did you use?" the man wanted to know.

"It's a keyframe animation system," Lasseter said. "You know, pretty much like what everyone else has."

The man persisted. "No, no, no. It was so funny. What software did you use?"—apparently in the belief that the software had built-in humor generation.

As with the Genesis-effect scene two years earlier, the defining qualities of *André and Wally B.* were to be found in the film's subtleties. One was that the characters had realistic motion blur; the new algorithm had worked. Another was that Lasseter had, for the first time in a computer-animated film, applied the classic animation principles that had evolved over the decades at the Disney studios. They were tools that animators used to move characters in a natural-looking way, to give characters an appealing look, and, above all, to make characters *act.* More often than not, the differences came down to a few seconds or even a fraction of a second of film.

For example, when Wally decides to chase the boy, he stares hard for a moment in the boy's direction before zipping off—as Lasseter followed the classic principle calling for anticipation (in whatever form) before action. Because the viewer sees Wally deciding what to do and preparing to do it, he or she perceives the ensuing action as more lifelike. Moreover, once Wally starts moving, his appendages follow at different speeds depending on their weight, with his heavy feet dragging behind more than his lightweight antennae—a touch of cartoon-style realism grounded in the classic animation principle of follow-through.

Lucas was polite in his reaction to the film. Only later did Catmull and Smith learn his real opinion: Privately, Lucas thought the film was awful. The character designs were primitive. The story was thin. The screening reinforced his feeling that his Computer Division shouldn't be making films, and it gave him a low impression of computer animation.

"He couldn't make the leap from the crudeness of it then to what it could be," Smith said. "He took it literally for what it was, and assumed that's all we could do."

Another filmmaking giant who was deeply skeptical of computer animation at this time was Frank Thomas, one of the so-called Nine Old Men—the elite of Disney's animators and directors from the 1930s through the 1970s. His book *Disney Animation: The Illusion of Life* (1981), co-authored with his friend and fellow "old man" Ollie Johnston, was and is the bible of technique for animators working in the Disney style. During the year or so before the premiere of *André and Wally B.,* the seventy-one-year-old Thomas, who was retired, had been visiting computer graphics researchers and animators at production houses and universities. He and Johnston had visited Lasseter while *André and Wally B.* was underway and, like Lucas, had been politely complimentary.

Shortly afterward, however, Thomas wrote a lengthy essay in which he seemed to argue that computer animation was a dead end. "Even today there is no electronic process that produces anything close to 'Snow White quality,' " he wrote, "and there is little reason to believe there ever will be. . . . This kind of animation may simply not be suited to this new medium."

He noted that the animation of characters with believable movements and believable emotions required tremendous subtlety. He was impressed by the "ease with which 3-D figures in 3-D surroundings can be generated" by a computer. He conceded that many of the classic animation principles could be carried over to the new medium. Yet Disney animators, he said, had learned that character animation required nuance on top of nuance—nuances too fine to capture in a computer:

> Today's computers can generate cartoon actors with rich personalities and put them in story situations that achieve full audience involvement. Weight and convincing movements are not too big a problem, and many facets of acting are within their capabilities. It is even possible to make the computer figures appear to think, but there it ends. The subtle pantomime, believable dialogue, appealing drawings, and most of all that personal artistic statement may be beyond our reach in the mechanical area of electronic circuitry.

The goal of computer animation made little sense in the first place, Thomas argued. It forced artists to go through endless convolutions to create their art. "It is so much easier to pick up a pencil and simply draw the action," he observed. "Old-fashioned animation has more control and more freedom, and also offers a greater range of expression."

Although Thomas appeared to be writing off computer animation in general, he came across at times as if he were rejecting only the idea of computerized 2-D cel animation. The ambiguity perhaps reflected the tentative, grasping nature of his own thoughts as he sought to understand the relationship between the new technology and the work he had done for a lifetime. Other traditional animators would be struggling with the same issues in years to come.

4. STEVE JOBS

By early 1985, it was obvious to Ed Catmull that the computer graphics group would soon need another home. Several storms were converging on the group and reinforcing one another: George Lucas's divorce had led to an expensive property settlement with Marcia, depleting Lucasfilm's cash reserves. Lucas had hired a new president, Doug Norby, with a mission to make the company's various pieces start functioning as viable businesses. Norby was already in the process of spinning off Andy Moorer's digital sound group and Ralph Guggenheim's digital film editor group into a company called The Droid Works. Computer graphics could not be far behind. Most important of all, Lucas and Catmull had fundamentally incompatible visions for the group.

The medium of motion pictures had had its start in the Bay Area a little more than a century earlier at an estate called Palo Alto, the future site of Stanford University. There, an engineer named John Isaacs helped the photographer Eadweard Muybridge capture still images of a horse and rider in motion. Central Pacific Railroad president Leland Stanford had financed the work to settle the question of whether all four of a horse's hooves ever leave the ground at once while it is in a gallop. (They do.) The photos, created with a bank of a dozen cameras, would make Muybridge a celebrity; Isaacs, whose electrically triggered shutters were the breakthrough that made the photos possible, remains an obscure figure. The trouble at Lucasfilm was that Lucas wanted his computer graphics experts to be Isaacs, the inventor, while *they* wanted to be Muybridge, the artist—or, more precisely, they wanted to be both Muybridge and Isaacs, with Lucas in the role of Leland Stanford writing the checks.

Lucas wanted the graphics group to make film production tools, not computer-animated cartoons.

Catmull and Smith talked fretfully about their concerns that Lucas might wake up one morning and decide to disperse their world-class team simply because he didn't know what to do with it. What actually happened was far more benign: Lucas's executives ordered Catmull to find a buyer and bring them the offers.

The graphics group was to be packaged as a computer hardware company. To carry out the task for which Lucas had created the group in the first place—building a set of devices that could scan movie film, combine special-effects images with live-action footage in a computer, and record the results back onto film—the group had been working on a specialized computer to sit in the middle of the process. Besides digital compositing, the computer was also expected to handle touch-ups, color correction, and rotation of frames to remedy camera misalignments. To perform such tasks on cinema-quality images would take enormous computer power by the standards of the day. Designed by an Evans & Sutherland alumnus named Rodney Stock, the machine was to process images as fast as a supercomputer, but at a fraction of the cost.

The Adventures of André and Wally B. had been the public face of the Lucasfilm Computer Division at the 1984 SIGGRAPH; privately, Lucasfilm had shown a prototype of the computer to a small audience in a hotel suite. In the eyes of Lucasfilm executives, this machine—not computer animation, and certainly not a Disney-trained "interface designer" named Lasseter—was the crown jewel of the computer graphics group. This was how Lucasfilm would get its cash out: Some company would buy the group, mass-produce the machine, and market it to industries that dealt with high-resolution images.

As befitted a computer that had been developed with money from science-fiction films, Lucasfilm's machine had an aura of science fiction itself, incorporating features that would not appear in mainstream computers for a decade or more. At the suggestion of Loren Carpenter, Stock had designed it with four processors and four banks of memory—one for each of the three primary colors and one left over. The fourth processor and memory bank would control the degree of transparency at any given point on an image. The processors ran in parallel so that a program could carry out an operation on all four parts of a single pixel (red, green, blue, and transparency) at the same time. Several of these four-processor combinations could be installed in one machine, so that up to twelve processors could be working at once.

During the planning stages of the machine several years earlier, the group had struggled to agree on a name for it. Finally, a foursome from the Computer Division had headed one evening in 1981 to a burger place—the Country Garden Restaurant in Novato, across the freeway from where they were based at the time—resolved to arrive at a decision over dinner. Someone brought up the idea of "Picture Maker." Smith argued for deriving a name, somehow, from the word *laser.* The group's film scanner and recorder used lasers, and he also thought the word sounded cool. He suggested "Pixer," a made-up word in the style of a Spanish verb. *Pixer:* to make pictures. Carpenter spoke in favor of a variation, Pix*ar.* The others around the table—Smith, Rodney Stock, and Jim Blinn (briefly away from his longtime professional home at JPL)—gave their assent. The machine would be called the Pixar Image Computer.

Catmull and Smith felt that if the group had to be spun off, it was best for the organization to become a hardware company—not that they knew much about the business side of computers. They were thinking like aspiring filmmakers, not businessmen, and their main goal was simply to keep the brain trust together. From their point of view, the wonderful thing about the computer hardware business was that computer companies needed a lot of employees. The computer graphics group was now up to forty or so people, and every one of them could be given a perch in a computer manufacturing company—somewhere or other. The group could then bide its time waiting for computer-animated feature films to become economically feasible.

Lucasfilm shopped the organization to twenty venture capitalists and investment banks, but found no takers. A series of manufacturing companies showed interest in a deal: The Dutch conglomerate Siemens saw the Pixar Image Computer as a complement to its CAT scanners, one that would allow doctors to look at scans in high-quality 3-D views for the first time. Several makers of color printing and scanning equipment—Scitex, Hell Graphics Systems, and Crossfield—were interested in selling the machine for prepress work. Perhaps on account of Lucasfilm's fifteen-million-dollar asking price (plus another fifteen million to capitalize the company), none of these would-be buyers stayed around for long.

Catmull and Smith visited the Kansas City, Missouri, headquarters of Hallmark Cards, Inc., which had responded to Lucasfilm's pitch. On the surface, the combination seemed incongruous; the greeting-card company's major acquisition the year before, Binney & Smith, sold Crayola crayons and Silly Putty. The idea was that Pixar Image Computers could fit in the pipeline between Hallmark's vast warehouse of hundreds of artists in cubicles, on one hand, and the company's sophisticated five- and six-color printers, on the other. The two men were enthusiastic about the prospect—they liked the people there, and the company at least did something with *art.* To their chagrin, Hallmark, too, took itself out of contention.

As the process wore on, Smith told Alan Kay what was happening and asked Kay to let him know if he had any ideas. Smith and Kay knew each other from the days when both were at Xerox PARC—Smith as digital artist, Kay as head of PARC's Learning Research Group. The two had not worked together there, but Kay regarded Smith as a good guy, and smart. Kay also knew Catmull from the

University of Utah. Not long after the conversation, Kay thought of a thirty-year-old multimillionaire of his acquaintance who he believed might be interested.

The friendship of Alan Kay and Steve Jobs originated with a computer called the Alto, a machine that Xerox PARC researchers conceived as a next-generation system for office workers. The Alto's user interface, created by Kay, was revolutionary: It could display pictures and text together on one screen; there was a device called a "mouse" with which the user could point to places on the screen and make the computer do things. There were programs to edit documents and graphics in such a way that the user would see on the screen exactly what they would look like when printed. Jobs had seen demonstrations of the system in December 1979 and had been thunderstruck.

"It was one of those sort of apocalyptic moments," Jobs reflected years afterward. "I remember within ten minutes of seeing the graphical user interface stuff, just knowing that every computer would work this way someday; it was so obvious once you saw it."

Later, when Kay was head of research at Atari, he would invite Jobs over for lunches, to the consternation of other Atari executives. When Apple Computer released the $2,495 Macintosh in January 1984, it turned out to be crafted in the image of a stripped-down Alto, combined with innovations from Apple's own designers. Jobs—who had obsessed over the Macintosh's design details—gave Kay a Macintosh to recognize his contributions to it. Jobs hired Kay as an "Apple Fellow" that May.

A year later, the situation had turned upside down: Kay was still working at Apple, but Jobs was gone. While many at Apple admired Jobs's vision and perfectionism, he had an obnoxious side that antagonized employees up and down the ranks. Jef Raskin, originator of the Macintosh project, made an eleven-point list of the reasons Jobs was impossible to work with and later resigned from the company. (Item 3: "Does not give credit where due." Item 4: "Often reacts *ad hominem.*" Item 10: "Often irresponsible and inconsiderate.") Jobs was in the habit of parking his Mercedes in a handicapped-only space in the front of the Macintosh team's building—because, it was said, if he parked in one of the regular spaces to the back or the side, someone was sure to sneak around and key his car.

John Sculley, the chief executive Jobs himself had recruited from PepsiCo, decided in the spring of 1985 to cut back Jobs's responsibilities. Sculley defined a new role in which Jobs would do nothing but envision new products; he would no longer be in charge of the Macintosh.

Jobs, incensed, plotted a boardroom coup to take place while Sculley would be traveling in China. Tipped off by one of his vice presidents at the last minute, Sculley canceled the trip. On Friday, May 31, 1985, Scully stripped Jobs of his position as executive vice president and general manager of the Macintosh division. Jobs would keep only the role of board chairman, largely ceremonial. His plot had thoroughly backfired.

Jobs found himself with time on his hands. He decided to take a walk.

He and Kay went for lunch at a healthy-food place, then hiked alongside CalTrans railroad tracks not far from Stanford. The conversation meandered. Kay mentioned his friends at Lucasfilm. "I told him that this really great group was looking to be loose, and that many things could be done with what they already knew and were interested in finding out," Kay recalled.

Soon Catmull and Smith were in Jobs's Spanish Colonial Revival mansion in the Silicon Valley town of Woodside. Friends who had worked with Jobs at Apple had warned them that Jobs would be charming during the meeting—but that if they were to sign on with him, he would drive them to distraction. Nonetheless, they trooped over from Kerner Boulevard to hear him out. They were joined by Ajit Gill of Lucasfilm's business development department, who was there to field any discussion of finances.

Woodside was a tastefully wealthy enclave that had retained a rural flavor; among its residents were venture capitalists and electronics industry executives, as well as the folk singer Joan Baez and Koko, the sign-language-talking gorilla. Jobs, a bachelor with an estimated net worth of some $185 million, lived a home life of extravagant asceticism: His seventeen-thousand-square-foot, fourteen-bedroom residence, built in the 1920s by a copper tycoon, was almost entirely empty. There was a BMW motorcycle parked inside, and a grand piano, and that was about it. Only the kitchen seemed to be fully outfitted; it was the domain of the husband-and-wife chefs, formerly with Chez Panisse in Berkeley, whom Jobs had hired to prepare his vegetarian meals.

He had bought the house the previous November, when he was still the Valley's golden boy. That same month, he and Sculley had appeared together in shirtsleeves on the cover of *BusinessWeek,* all smiles, beneath the bold headline "Apple's Dynamic Duo." Now, suddenly, his professional life was as empty as his house.

Catmull, Smith, and Gill spent the afternoon with Jobs as he alternated between complaining angrily about Sculley and talking about his ideas for the Computer Division. Finally he took them for a walk outside and proposed to buy the group and run it. Catmull and Smith told him they weren't interested; they would be happy to take his money as an investment, they said, but they wanted to run the operation themselves. They parted amicably. Jobs told them he thought Lucasfilm's asking price was too high and asked them to let him know if it came down. Five million would be more like it, he said.

Afterward, Catmull told Smith it was the wrong time to become involved with Jobs. Jobs was still too emotional over Apple, too bitter. Catmull had been through a divorce with his first wife, Laraine, shortly after coming to Lucasfilm; he saw the signs. "We don't want to be the first woman after the divorce," he concluded.

The group continued to pursue several film-related projects. One of these was for Steven Spielberg, Lucas's friend and sometime partner, who had been impressed with the Genesis effect. Spielberg was producing the Barry Levinson–directed film *Young Sherlock Holmes,* which called for a hallucination scene in which a knight in a stained-glass window would seem to come alive and terrify an elderly priest. Lasseter designed and animated the knight. It was the first use of the

Pixar Image Computer and David DiFrancesco's laser film printer in a feature film; it was also the first time a computer-generated character had been combined with live-action scenery.

After finishing work on *Young Sherlock Holmes,* Lasseter launched an experiment. He was well practiced on the group's animation program, Motion Doctor, thanks to *André and Wally B.* and the stained-glass knight. Others on the staff, however, had always constructed the 3-D models for him under his direction. He decided to try an exercise with Bill Reeves's modeling program, ME (for Model Editor), to expose himself to the art of model making.

He cast around for something to build and his eyes settled on the lamp on his drawing table, a Luxo. He measured the lamp with a ruler and worked out the geometric shapes it would need.

The group had no interactive modeling tools—there was no way to rotate *this* part and stretch *that* part while watching the object change on the screen. Instead, ME required the model maker to use a modeling language, a kind of programming language. Lasseter would type lines of code using a text-editing program to define the way the object was put together, save them in a computer file, and then watch ME draw a wire-frame version of the object. If it didn't look right, he would have to go back, try to fix the file, and repeat the cycle.

Once he finished the lamp, he found he wasn't quite ready for the experiment to be over. He thought: *Now that I've made this thing, let's see if I can make it do something interesting. Something . . .*

Catmull and Smith, meanwhile, were pursuing a potential collaboration with a large Japanese publisher, Shogakukan. The heir apparent of the company wanted to make a feature film based on the beloved fantasy-hero character "Monkey" of Chinese and Japanese mythology. Monkey was the star of the sixteenth-century Chinese novel *The Journey to the West,* one of the classical works of Chinese literature. The story, derived from ancient legends, was an unusual blend of spiritualism, comedy, and adventure: It told of Monkey, an accomplished trickster and magician, accompanying a priest who trekked from China to India to obtain the holy texts of Buddhism for the Orient.

The Shogakukan team judged that a computer-animated film of the Monkey story would bring tremendous prestige. They expressed enthusiasm for the ideas coming out of the computer graphics group; there were many meetings, including a weeklong brainstorming session at a resort in the California beach town of Carmel. Catmull and Smith were excited that the first feature-length computer animation might be within their grasp. At the same time, they were concerned that neither side had any real understanding of film marketing or distribution. Smith wondered whether the heir apparent was overreaching and whether the film's exorbitant cost—fifty million dollars or more—might prove embarrassing to him. Nothing could be decided conclusively, in any case, until the spin-off was completed.

At the same time, the group had still another effort underway, namely, negotiating with Walt Disney Productions on a major contract for consulting services and hardware. Catmull and Smith had kept up their annual visits to the company, begun during the New York Institute of Technology years. Now, finally, senior management at Disney appeared to be serious about putting computers to work on traditional 2-D animation.

In the early 1980s, Catmull and Smith had persuaded a Disney engineer named Lem Davis of the merits of using computers to replace the battalions of women who laboriously painted cels by hand in the Ink and Paint Building. With Davis's support in hand, they thought, *Great, Disney's finally going to do something with computers in animation.* Prematurely, they assumed that the door was unlocked and swinging open.

Their reaction was a typical neophyte's error; in a company like Disney, winning the support of the technical guys was only the first and easiest tumbler in the lock. Every layer of management above would have its own priorities and agenda.

Davis eventually persuaded Stan Kinsey, senior vice president of operations and new technologies at Walt Disney Studios, that Catmull's and Smith's ideas were just what Disney needed: They would not only lower costs, but also allow freer camera moves and a richer use of colors. Computers could bring Disney animation back to the level of vibrancy and spectacle it had achieved in the 1930s and 1940s. Kinsey, however, was unable to win approval to move the project ahead.

The concept seemed doomed until Saturday, September 22, 1984—the morning that Disney's board voted to hire former Paramount Pictures president Michael Eisner as CEO and former Warner Bros. president Frank Wells as president and chief operating officer. They were there thanks to a coup led by Walt's nephew, Roy E. Disney, and Roy's business partner, Stanley Gold. Over lunch that afternoon at Lakeside Golf Club in Burbank, Eisner and Wells asked Roy, their benefactor, what he would like to do. He hadn't considered it. "Why not let me have animation?" he suggested off the top of his head.

"Great," Eisner replied.

Eisner and Wells had been inclined to close the moribund animation division. It was simply too far gone. If it would keep Roy Disney happy, though, it was a worthwhile price to pay.

Kinsey found an ally in Roy, who was excited by a briefing that he received on computer animation from Disney engineers that fall. The briefing, scheduled for forty-five minutes, ended up taking twice that long as Roy asked questions and requested repeated showings of *André and Wally B.* and the test footage of *Where the Wild Things Are.* Within a month after the new regime took office, Roy and Kinsey were in touch with Catmull. Teams from both companies began mapping out the specifics of the project: the Computer Animation Production System, or CAPS. It would use the Pixar Image Computer and custom software to automate the inking and painting of an animator's pencil drawings.

Kinsey wanted to go further and buy the graphics division from Lucasfilm outright, believing that its fifteen-million-dollar price was a bargain. Jeffrey Katzenberg, a protégé of Eisner who now oversaw all Disney filmmaking as head of Walt Disney Studios, overruled this idea. "I can't waste my time on this stuff," Kinsey remembered Katzenberg telling him. "We've got more important things to do."

A consulting deal would be fine, however. Yet Catmull and Smith still weren't at the finish line: Under Eisner, Walt Disney Productions was already earning a reputation as hard-nosed and difficult to do business with. The discussions of business terms moved slowly and continued through 1985.

The constant quest for a buyer hung over everything the group did. Smith felt drained by endless battles with Lucasfilm executives who wanted to cut the size of the group—making it a more attractive package, they believed. The lower the head count, the lower the operating costs, the better the financial story. Smith was convinced that they had the situation upside down: The real value wasn't in the hardware designs and the software; it was in the collection of extraordinary talent. "The whole year was just this torture of protecting our people from these supposed businesspeople," Smith remembered.

In November 1985, the group had what seemed to be a last-ditch opportunity to close a deal. The buyers were the Dutch colossus Philips Electronics and the Electronic Data Systems subsidiary of General Motors. Philips was interested for much the same reason Siemens had been—it planned to sell the Pixar Image Computer as an adjunct to its MRI scanners. Catmull and Smith had persuaded management at EDS and GM that the group could provide advanced hardware and software for automobile design with highly realistic rendering. As it happened, the auto industry had been the source of early breakthroughs in the mathematics of modeling 3-D surfaces, most notably in the work of Pierre Bézier, an engineer at Renault, in the early 1960s. GM itself had been a computer graphics pioneer during the same period, when a team within GM Research Labs developed one of the first computer-aided design and manufacturing systems. (GM management at the time did not appreciate the value of what it had and treated the system as an experimental toy before abandoning it altogether.)

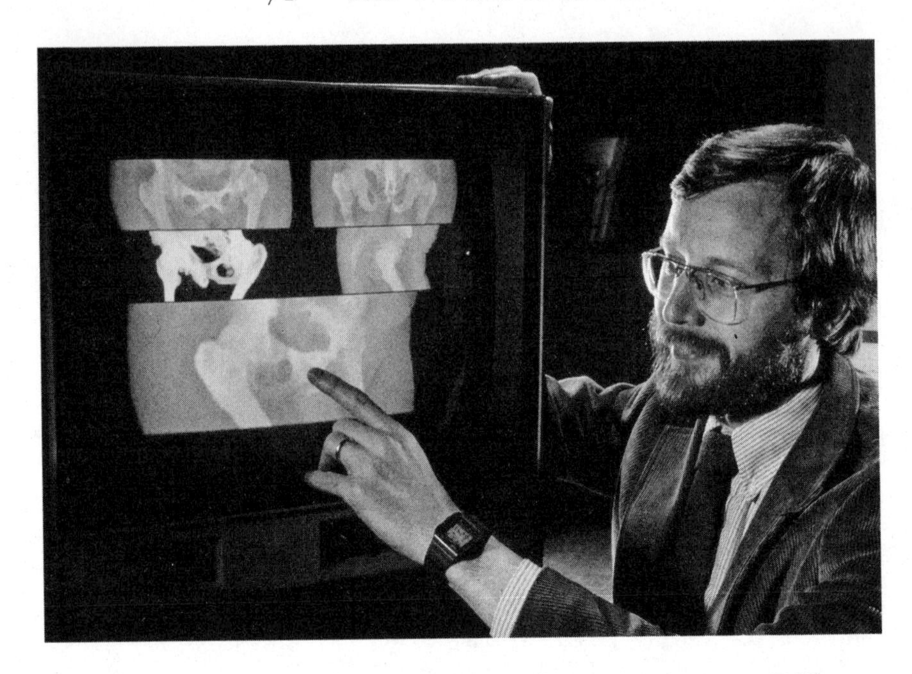

Ed Catmull, at Lucasfilm in 1985, showing off a medical image—a CAT scan of a woman's pelvis—assembled into 3-D form by a prototype Pixar Image Computer. Catmull and Alvy Ray Smith tried to arrange the sale of the Computer Division to the medical electronics companies Philips Electronics and Siemens.
© Roger Ressmeyer/CORBIS

Under the terms that the companies had worked out, Lucasfilm was to retain one-third ownership of the spin-off company, while Philips and EDS would each pay $5 million for one-third shares. Another $4 million would be raised from venture capital firms. Of the $14 million total, $11.5 million would go to Lucasfilm and $2.5 million would capitalize the new company. Philips and EDS would get exclusive rights to sell the Pixar Image Computer in their markets. Catmull and Smith, together with representatives of Philips, EDS, and Lucasfilm, spent two days in a Philips boardroom high above Forty-second Street in New York negotiating the details. At the end of the second day, everyone present agreed that they had a deal.

Or so they thought.

By coincidence, GM's board met the next morning in the General Motors Building about a mile uptown to vote on a proposed $5.2 billion buyout of Hughes Aircraft, a satellite and missile contractor. Ross Perot, EDS founder and GM board member, dissented from the buyout and used the occasion to tell off everyone in the room. GM executives seemed to care only about procedures, he said, not results. Its directors were passive time servers who seemed to forget their job was representing shareholders. GM cars had a problem with reliability and no one was doing anything about it, he charged. He had visited a conference of Cadillac dealers and they were in an uproar about the rampant defects in the cars. How, Perot asked, could GM justify spending billions on a communications satellite operation when it couldn't even build a reliable car?

The company's chairman, Roger Smith, thanked Perot for his comments and moved on.

Within General Motors, any deal with EDS attached to it was now radioactive. There would be no investment in the graphics division. Moreover, Philips either could not or would not proceed on its own.

In the meantime, Jobs had resigned his Apple chairmanship on September 17, severing his last tie with the company apart from his stock ownership, and had taken five employees with him to start a new computer company, Next (later NeXT). Throughout the year, he had been checking in with Lucasfilm to see whether the Computer Division's price had come down. Now, ten months after putting the organization on the market, Lucasfilm seemed to have no alternatives left. Jobs surmised as much.

Shortly before Christmas, Jobs phoned Lucasfilm president Doug Norby, who, by now, was eager to wash his hands of the graphics group any way he could. The sale process had been a major distraction and the group's forty-person payroll was a money sink. If he couldn't sell it by the end of the year, he had decided, he would shut it down—thank you, and good-bye. Jobs offered him five million dollars, precisely the figure he had put forward at the outset. Norby accepted.

At the last minute, after the fine points of the deal had been worked out, a new issue cropped up that threatened to scuttle the transaction: Where, exactly, would the parties sign the documents? Lucasfilm vice president Doug Johnson thought it made sense for Jobs to come to Skywalker Ranch in the hills of Nicasio, California; Jobs maintained that Lucasfilm should come to *his* office at NeXT in Redwood City, sixty miles away. Neither side was willing to budge. Finally, they settled on the San Francisco offices of Lucasfilm's law firm, Farella, Braun & Martel, as a halfway point.

On Thursday, January 30, 1986, Catmull, Smith, and Johnson signed papers creating a new company and turning the graphics group and its technology over to the company. The spin-off was named Pixar, Inc., for the computer. On February 3, the following Monday, the three of them sat down with Jobs in a law firm conference room to sign the papers transferring ownership of the Pixar stock. Catmull and Smith, as co-founders, would each own 4 percent; another block of stock, vesting over a period of time, would be spread among the other thirty-eight employees. Jobs brought with him a five-million-dollar cashier's check for his stake, plus a one-million-dollar cashier's check as the first installment of the five million in capital he had agreed to put into the company.

Jobs enjoyed a reputation as a visionary reader of consumer markets, a reputation he had earned time and again. Yet if he had possessed the same eye for reading individual human beings, he might have noted something unsettling about the men he had just taken on. He might have observed that Catmull and Smith—the chief technical officer and the vice president, respectively, of his new computer hardware company—had no particular interest in computer hardware. For them, computers were simply boxes, the means to an end. If it made the most sense to get a job done with a Sun, or a VAX, or a borrowed Cray, that was the machine they used. For Jobs, a well-designed computer could have the allure of a sports car. (The Macintosh should be "like a Porsche," he had pronounced during a design meeting.)

The differences went deeper still. In their computer graphics work, Catmull and Smith were in pursuit of tools for great artists; the group's rendering and animation programs were akin to Stradivarius instruments—fit for a world-class orchestra, wasted on amateurs. The notion that computers could be an empowering force in the hands of ordinary people was neither right nor wrong so much as it was simply beside the point. As Catmull and Smith saw things, personal computers like the IBM PC and the Macintosh were playthings, not to be taken seriously. Neither man used one; they had far more powerful systems at their disposal.

Indeed, when David DiFrancesco had urged Smith to visit one of the ILM buildings next door on Kerner Boulevard to see what an ILM programmer and his brother were doing with image processing on the Macintosh, Smith never got around to it. "I couldn't be bothered with tiny little machines at the time," Smith recalled later. (The brothers, Tom and John Knoll, later licensed their software to Adobe Systems; it went on to fame under the name Photoshop.)

Jobs cared very much about the "tiny little machines." They had been his life.

Steven Paul Jobs was born in San Francisco on February 24, 1955, and adopted as an infant by Paul and Clara Jobs. Paul, who had never finished high school, worked variously as a debt collector and a machinist. When their son was five, the Jobses moved from San Francisco to the Silicon Valley city of Mountain View. There, Paul kept a workbench in the garage. One day, he brought his son out to the garage, gave him a hammer and saw and some other tools, marked off part of the bench, and told him, "Steve, this is your workbench now."

Paul spent a lot of time with him in the garage, showing him how to build things.

A little later, Steve got to know a neighbor down the street, a Hewlett-Packard engineer named Larry Lang, who taught him some electronics and introduced him to Heathkits. Heathkits were everyday electronics products—a battery tester, a transistor radio, an amplifier—that came unassembled. A manual showed how to put the components together and explained the theory of how the device worked. The purchaser actually paid more for a Heathkit than for a finished version of the same item, but it also provided something that the off-the-shelf product didn't.

"It gave one the sense that one could build things that one saw around oneself in the universe," Jobs remembered.

> These things were not mysteries anymore. I mean you looked at a television set, you would think that, "I haven't built one of those, but I could. There's one of those in the Heathkit catalog and I've built two other Heathkits so I could build that." Things became much more clear that they were the results of human creation, not these magical things that just appeared in one's environment. . . . It gave a tremendous level of self-confidence, that through exploration and learning one could understand seemingly very complex things. . . .

Jobs came to hold an intense dislike for his elementary school, which, by the standards of the Bay Area, drew from some rough neighborhoods and had problems with disorder and indiscipline. One day he announced to his parents that if they didn't find another school for him by the next school year, he would stop going to school. They realized he meant it. The Jobses moved again, this time to the town of Los Altos a couple of miles away, drawn by the well-regarded public schools of the Cupertino school district.

As a freshman at Homestead High School, he enrolled in the school's Electronics 1 class. One evening, in need of parts for a class project, he phoned Bill Hewlett, co-founder of Hewlett-Packard, at home and asked whether Hewlett-Packard could spare them. Jobs got the parts and a summer job on an HP assembly line.

During the summer before Jobs's senior year, a school pal of his, Bill Fernandez, introduced him to an older friend from Fernandez's neighborhood. The friend's name was Steve Wozniak, universally known as Woz. Fernandez recalled that he and Wozniak were "electronics buddies" who worked on projects together. They had just collaborated on building a computer of Wozniak's own design; Fernandez invited Jobs to drop by to see their finished work. Fernandez reckoned that since Jobs and Wozniak were both interested in electronics, they might like to meet each other. Jobs had honed his knowledge of electronics over the years, but he quickly recognized that the twenty-year-old Wozniak was on another level entirely.

The two became friends. Wozniak, the son of a Lockheed engineer, was a self-taught expert on computer design. He had studied electrical engineering for several semesters at the University of Colorado and De Anza Community College, and had taken all of the three years of electronics instruction offered at Homestead during his high school days; his real education, however, had come from reading the circuitry manuals and schematics for minicomputers from Digital Equipment Corp. and Data General and working out his own designs.

That fall, Wozniak read an article in *Esquire* magazine about a subculture of eccentrics, the "phone phreaks," who were able to call anywhere in the world for free by replicating the master tones that controlled the phone system. The AT&T telephone monopoly had unwisely allowed one of its technical journals to publish the necessary frequencies as part of an engineering article, putting the information into libraries across the country. Now an underground movement was devoted to building and using highly illegal "blue boxes," devices that offered phone privileges without limit. For many, the attraction was psychological as much as monetary. "This huge thing is there. This whole system," one of the phreaks told *Esquire.* "And there are holes in it and you slip into them like Alice and you're pretending you're doing something you're actually not, or at least it's no longer you that's doing what you thought you were doing. It's all Lewis Carroll."

Wozniak, excited, called Jobs before he had even finished reading. The two had not been interested in the telephone system up to then, but the challenge was irresistible. The next day, they found the frequencies in the library of the Stanford Linear Accelerator Center and set to work. They tried at first to build a garden-variety analog blue box with oscillator circuits, but the oscillators proved erratic. The answer, Wozniak decided, was a digital blue box—more difficult to design, but easier to use.

At Jobs's instigation, they turned the digital blue box into their first business venture, peddling it in male dormitories at Berkeley. Wozniak refined the design until the user didn't even need to press an on–off switch; pressing any of the buttons on the keypad started the machine. The parts cost about $40 and the unit sold for $150; they sold around $30,000 worth. "He wanted money," Wozniak later explained.

Jobs graduated from Homestead in 1972. AT&T's rights in the matter of the blue box did not weigh on his mind, but the danger of getting caught did. He ended his involvement in the business soon afterward. Nonetheless, he'd had his first taste of entrepreneurial success.

Jobs had visited a friend at Reed College in Portland, Oregon, that year and felt inspired by its Bohemian ambience and Pacific Northwest setting. Upon his return, he announced to his parents that he wanted to go to Reed. Paul and Clara were in a bind: When they were seeking to adopt him seventeen years earlier, his birth mother, a graduate student, had changed her mind about signing the adoption papers after learning that Clara never graduated from college and Paul was a high school dropout. They overcame her reservations only after pledging that Steve would go to college.

In all the years since, Paul and Clara had never gotten out of the lower middle class. They tried to talk Steve out of the small, expensive private school, but he was unbending. It was Reed or nowhere, he said, and so they moved him into a Reed dormitory that fall.

The school turned out to have high academic expectations, and its great-books curriculum was not to Jobs's liking. He dropped out at the end of 1972. "After six months, I couldn't see the value in it," he later said of the experience. "I had no idea what I wanted to do with my life and no idea how college was going to help me figure it out."

He hung out in Portland for a couple of years, living off a job maintaining electronic equipment for the school's psychology department. He came back to his parents' house in 1974 to find more lucrative work in the Valley, hoping to save money for a trip to India with a college friend. Thanks to a classified ad in what was then the *San Jose Mercury,* he found a job as a five-dollar-an-hour technician at the video game company Atari.

The company was in its second year. Despite Jobs's low-ranking status, he made a habit of informing Atari engineers that they were moronic and their designs were lousy. This, combined with his unconventional hygiene—he had become convinced in Portland that a "fruitarian" diet of raw fruit and seeds obviated the need for bathing—led his superior to put him on a night shift where he could work alone.

Over the next year, Jobs bounded from Atari to India (he persuaded Atari to subsidize his airfare as far as Germany), to primal-scream therapy in Eugene, Oregon, to Atari again for a while, to a commune-style apple farm, and finally back to Atari.

Jobs returned to the Bay Area just as it was becoming one of the centers of a nascent cultural revolution—one as significant, in its way, as the one that the Bay Area had helped to bring to life in the mid-1960s. In 1975, one could find pieces of the revolution in wildly disparate parts of America. In Albuquerque, a calculator company called MITS was fitfully rolling out the $397 Altair 8800, the first low-cost personal computer that could be considered more than a toy. In a two-bedroom apartment in the same city, three Harvard undergraduates, Bill Gates, Paul Allen, and Monte Davidoff, were working on versions of the BASIC programming language for the machine. The town of Peterborough, New Hampshire, was the home of a new glossy magazine, *BYTE,* that catered to a vanguard of personal computer hobbyists; editor Carl Helmers asked invitingly in the first issue, "Wouldn't it be neat to have a computer all one's own without being as rich as Croesus?" In Jobs's environs, a club devoted to small computers was beginning to meet every two weeks—first in the garage of one of the founders, later at a private school, and finally, as the crowds grew, at the Stanford Linear Accelerator Center auditorium.

It was called the Homebrew Computer Club. Wozniak started attending before Jobs. Attendees would hear reports from others who had built something new or written some interesting software. After the regular session, people would trade schematics and pirated software and answer one another's questions. "The theme of the club was, 'Give to help others,' " Wozniak remembered.

There, Wozniak and Jobs learned about the new generation of processors on a chip, the Intel 8080 and Motorola 6800. Wozniak felt encouraged to try designing his own machine around a processor chip and found a cheaper one, the MOS Technology MC6502. His computer would differ radically from those coming out of MITS and a few other small companies. The Altair and its rivals generally had a front panel with switches and blinking lights; to do anything useful with them, one had to buy an interface board to connect the computer to a Teletype machine. Wozniak designed a computer that connected to a keyboard and a video monitor or TV set—no extra interface board, no slow, noisy Teletype.

Wozniak freely shared schematics for the machine at club meetings and even went to people's houses to help them build their own. His reward was simply the respect of his peers in the club.

Jobs, however, had grown to venerate Atari's entrepreneurial founder, Nolan Bushnell. "Nolan was his idol," Wozniak said. "Steve wanted to have a successful product, go out and start selling it, and make some money."

As with the digital blue box, Jobs convinced Wozniak that the computer ought to be a product. Perhaps with thoughts of his apple-orchard days or his fruitarian diet in mind, Jobs suggested the name Apple for the machine and the company. To raise money for a professional to lay out a printed circuit board, he sold his Volkswagen Microbus and Wozniak sold his two HP calculators. They showed off the Apple I at the Homebrew Computer Club in mid-1976. An early computer retailer, Paul Terrell, was at the meeting; he noncommittally told Jobs that the product looked interesting and suggested that he keep in touch.

The next day, Jobs turned up at Terrell's store in Mountain View, the Byte Shop. "I'm keeping in touch," he announced.

Notwithstanding his stringy beard, cutoff jeans, and bare feet, Jobs could be persuasive. By the time Jobs left, Terrell had agreed to buy fifty Apple I's, fully assembled. Over the next month, Wozniak and Jobs scrambled to fill the order, and Jobs worked out the computer's suggested retail price: double the cost of materials, plus a 33 percent dealer markup—$666.66.

Apple would sell only a couple of hundred Apple I's in all, but it was the starting point of a computer empire. When the company had its initial public stock offering four and a half years later, in December 1980, it was the largest initial public offering since the Ford family took Ford Motor Company public in 1956. So rapid was Apple's ascent that in 1982, while Jobs was still in his twenties, *Time* magazine planned to name him its Man of the Year (following Ronald Reagan and Lech Walesa, the selectees for 1980 and 1981).* The editors substituted "The Computer" at the last minute because they concluded it was too risky to tie the designation to what was essentially a one-product company.

The design of the Apple I was wholly in Wozniak's hands. As Jobs put his imprint on subsequent Apple products—most notably the Apple II (1977), the LaserWriter (1985), and the Macintosh—what emerged was a synthesis of Apple's top-drawer engineering talent and Jobs's distinctive vision of computing. Jobs was a populist

* Jobs's growing reputation was briefly vulnerable to scandal when a former girlfriend gave birth to a daughter in 1978 and sought a modest amount of child support. Although a blood test indicated a 94 percent likelihood that Jobs was the father, he refused for two years to acknowledge paternity. Their friends, convinced that he was indeed the baby's father, could not understand his intransigence. Wozniak theorized that Jobs disagreed with his ex-girlfriend's decision to have the baby and became upset by his lack of control over the situation. Finally, in 1980, Jobs brought the matter to a close by agreeing to pay $385 a month in child support and to reimburse San Mateo County for the public-assistance money that the mother had received to help support the child while working as a waitress and housecleaner.

among computer entrepreneurs—not in the sense that Apple products were the lowest cost (they rarely were), but in his ambition to bring the high end of computing to the masses.

If there was a Steve Jobs formula, this was it. The Apple II took the Apple I as its starting point and added color graphics and an appliance-like feel. Where owners of the Apple I had to add their own keyboard, power supply, and case, the Apple II needed only to be hooked to a video screen and turned on. It offered computer power in an approachable form. It was a more egalitarian machine in *that* sense. (Indeed, its combination of functionality and approachability made it standard-issue in school classrooms.) With the Macintosh, Jobs brought a state-of-the-art, Xerox PARC–influenced graphical interface into the mainstream in an appealing package.

The LaserWriter, although costly at $6,995 when introduced, was nonetheless significant because it democratized PostScript, a computer language for complex page layout and typesetting. When Jobs learned that John Warnock and Charles Geschke were developing PostScript at their startup company, Adobe Systems, they were planning to use it only in a proprietary computer system with a twelve-hundred-dots-per-inch printer; it was Jobs who persuaded them to adapt PostScript to control lower-cost, three-hundred-dots-per-inch laser printers such as the LaserWriter. The Macintosh and the LaserWriter together became the foundation of desktop publishing, which would ultimately bring professional-style layout and typesetting within reach of ordinary users.

Jobs's practice of making high-end technology accessible to the mainstream served Apple's balance sheet, to be sure. Yet Jobs had more than purely commercial aspirations in mind; he had absorbed a new countercultural view of computers that had been percolating through the Homebrew Computer Club and a few other Bay Area institutions. While shaving his beard and trading his cutoffs for suspenders and a bow tie, he had remained in tune with the excitement surrounding this philosophical shift. "You talk to some of the people in the computer business now and they're very well grounded in the philosophical traditions of the last hundred years and sociological traditions of the sixties," he told an interviewer in 1984. "There's something going on here, there's something that's changing the world and this is the epicenter."

What was new was a notion among some with an anti-establishment cast of mind—who might otherwise have been disposed to regard computers as tools of authority—that *small* computers could be instruments of personal liberation. A quirky tabloid called *People's Computer Company,* published in Menlo Park, carried the message of "computer power to the people" amid printouts of games written in BASIC and drawings of dragons. In 1974, a Tom Paine–like propagandist named Ted Nelson self-published two books, bound together in a single set of covers: *Computer Lib: You Can and Must Understand Computers Now,* touting personal computers, and *Dream Machines: New Freedoms Through Computer Screens,* speculating on the potential of computer graphics. He sold thousands of copies. The front of *Computer Lib* was adorned only by a fist, clenched in the style of a protest marcher. "COMPUTERS BELONG TO ALL MANKIND," the book proclaimed inside.

Jobs espoused those values and gave every indication that he believed them. Apple, he never tired of repeating, was a place for people who wanted to change the world. "When Apple was incorporated," Fernandez remembered, "and I started going over to [Jobs's] house to work as an employee rather than to hang out or do electronics projects, there was, at least for me, a tangible feeling of magic in the air." The magic came, he said, from the group's belief that they were bringing " 'power to the people,' meaning to put the power of computing into the hands of everyone."

Wozniak later wrote earnestly, "Our first computers were born not out of greed or ego but in the revolutionary spirit of helping common people rise above the most powerful institutions."

The philosophy had perhaps its most visible expression in the company's famous "1984" commercial, which aired on Super Bowl Sunday, 1984, a little more than two years before Jobs's buyout of Pixar. The commercial showed a sea of slack-jawed drones watching Big Brother on a massive video display until an athletic woman

1-1-1-1

From: **Pixar**
P.O.Box 2009
San Rafael, CA 94912

Contact: **Cunningham Communication, Inc.**
1971 Landings Dr.
Mt. View, CA 94043
Andrea Cunningham
(415) 962-8914

For Immediate Release

STEVEN P. JOBS AND PIXAR EMPLOYEES BUY PIXAR

San Rafael, Ca., February 10, 1986 -- Pixar, the computer graphics division of Lucasfilm Ltd., announced today that it has been acquired by Steven P. Jobs and the employees of Pixar. Pixar, now an independent company, will design, manufacture and market high performance computers and software specifically tailored for state of the art computer graphics and image processing applications.

Mr. Jobs paid an undisclosed sum in the millions to Lucasfilm Ltd. for a majority interest in the firm. The employees own the remaining shares. Pixar's Board of Directors will be comprised of Jobs, chairman, Edwin Catmull, president of Pixar, and Alvy Ray Smith, vice president of Pixar.

The new firm has a product, the Pixar Image Computer, ready for market. Developed during the last three years at Lucasfilm Ltd., the Pixar Image Computer is over 200 times faster than conventional minicomputers at performing complex graphic and image computations. At these specialized tasks, the Pixar Image Computer is also faster than a $6 million supercomputer. The Pixar Image Computer will be introduced to the commercial and scientific markets within the next 90 days and will sell for approximately $125,000.

2-2-2-2

Pixar was originally formed in 1979 by George Lucas to bring high technology to the film industry. Lucasfilm Ltd. will continue to use the Pixar Image Computer and other technologies to produce computer animation for films through its special effects division, Industrial Light & Magic (ILM), and for home entertainment through its Games Group.

Mr. Catmull said, "Society's ability to generate large amounts of data far exceeds its ability to assimilate this data. High performance graphics holds great promise for analyzing and rapidly processing this data into a form which we can see and use. The Pixar Image Computer cost-effectively processes and displays large amounts of data in new and unique ways for the benefit of professionals in medicine, geophysics, printing, remote sensing and other industries."

Mr. Jobs said, "Image computing will explode during the next few years just as supercomputing has become a commercial reality within the last several years. The technology is just now ready and Pixar will be the first to define and pioneer this new segment of the computer industry. I'm excited to be associated with this emerging new field of technology."

Edwin Catmull (40) and Alvy Ray Smith (41) both hold PhD's in computer science and have been leaders in the field of computer graphics for over ten years. Steven P. Jobs, a young industrialist, is founder and president of NEXT, a silicon valley startup which will make powerful computers for the higher education market.

came from the back of the room and tossed a sledgehammer into the screen. The closing frames promised that the Macintosh would be the reason "why 1984 won't be like '1984.' "

After buying Pixar, Jobs was optimistic that history would repeat itself: Graphics computers would start in the hands of a few early adopters and then make their way into a vast mainstream market. "This whole thing has the same flavor as the personal computer industry in 1978," he told *BusinessWeek* at the time.

At the newly formed company, Ed Catmull and Alvy Ray Smith were considerably more earthbound. The Pixar Image Computer was a nice machine, certainly, but it wasn't a get-rich-quick scheme. They grasped that Jobs's ebullience had gotten ahead of reality. "We thought computer graphics *was* going to change everything," Smith recalled, "but we had a long way to go to get there."

Catmull and Smith, now faced with real profit-and-loss responsibility for the first time in their careers, set to work figuring out how to bring in enough revenue to keep a roof over their collection of talent.

5. PIXAR, INC.

Outwardly, little changed at Pixar's offices after the group left the Lucasfilm fold in early 1986. The organization remained in one of the five buildings of the ILM complex on Kerner Boulevard in San Rafael, cheek by jowl with George Lucas's special-effects artisans. To avoid intrusions from overeager Lucas fans, the bland white buildings had been marked only with the letters *A* through *E*—except that the front door on building C had a subterfuge sign left over from ILM: KERNER OPTICAL RESEARCH LAB. In the parking lot behind the buildings, Pixar employees might, on any given day, encounter a special-effects miniature being set on fire or blown up as ILM's cameras rolled.

Within Pixar, Jobs quickly ended the "Monkey" project. Pixar, after all, was a hardware company. Catmull and Smith felt it was too early for a feature film to be economically feasible, in any case. Smith also felt protective of the people from Shogakukan, the Japanese company backing the film, with whom he had formed a bond; he was concerned that the spiraling estimates of the cost would lead to a loss of face for the corporate heir apparent who had stuck his neck out. Jobs otherwise kept his hands off Pixar, for the most part; he was intensely preoccupied with his other venture, NeXT.

Catmull and Smith adjusted themselves to their new roles as managers of what was, in essence, a start-up company. They subscribed to *Inc.* magazine and bought some business books, including the primer *Buy Low, Sell High, Collect Early & Pay Late.* Seeking advice, they phoned their former colleague from NYIT, Jim Clark, who had since founded a highly successful computer company called Silicon Graphics. "Oh, it's pretty easy," he said vaguely. "You'll figure it out in about a year."

Although the company began life as Pixar, Inc., Catmull and Smith dropped the Inc. once they learned that it wasn't required. The name Pixar standing alone seemed strong and simple. Pixar Animation Studios, both the name and the underlying reality, was some years away.

Hiring was a new kind of challenge, even though they had engaged scores of people at Lucasfilm and NYIT. Hiring a graphics expert was comparatively easy because they knew how to gauge talent and they tended to know who all the best people were. For positions in such alien territory as finance and marketing, however, they had to rely on headhunters and hope for the best.

At the time of the spin-off, the Pixar Image Computer still existed only as a handful of prototypes. Jobs gave Catmull and Smith advice about manufacturing, but they were rarely able to use it because it came from Jobs's realm of mass-market products that sold in the hundreds of thousands of units. To oversee the computer's production, Catmull and Smith hired a vice president of manufacturing and engineering, Chuck Kolstad, thirty-nine, an MBA who had previously been a manufacturing executive at the telephone equipment firm Rolm. Buttoned-down and business-minded compared with most of Pixar's academic types, he nonetheless impressed them with his knowledge of hardware production issues. Kolstad, for his part, found the setting more collegial than at other companies where he had worked. Even by the standards of Silicon Valley at the time, moreover, he found the employees unusual in the extent to which they were more interested in research than money. "We had people there we'd have to remind to go cash their paychecks," he said.

Pixar's first product, the ill-fated Pixar Image Computer. The machine offered supercomputerlike performance when analyzing and enhancing images, but after an initial burst of sales, it found few customers. John Lasseter provided the dimple-in-square design of the front panel.

By mid-1986, Pixar Image Computers were emerging from an assembly room on the premises, where workers stuffed circuit boards into gray metal cases. The front panel was a John Lasseter design, a beveled square with a round dimple in the middle; it doubled as the company logo.

In theory, the machine had significant markets awaiting it, fields that dealt with large images and had money to spend: radiology, scientific research, oil exploration, and defense, among others. Early sales, including an order from Philips Electronics for dozens of units, gave Jobs reason to feel vindicated. The Pixar Image Computer, or PIC, quickly made its way into major universities. By August, Pixar had entered into remarketing agreements with Philips and two computer manufacturers, Sperry Corp. and Symbolics.

"Steve's vision was that we were going to populate the world with Image Computers," Kolstad recalled.

Yet there were clouds overhead from the outset. After spending

$125,000 on the unit, the customer still needed an expensive Sun workstation to attach to it because the PIC lacked its own user interface. The PIC also lacked software tailored to Pixar's hoped-for markets; for this, customers had to wait for independent software developers or else write their own. "Who's going to buy a $125,000 image processor that requires a host computer and has software development tools but no applications software?" wondered *Computer Graphics World,* an industry magazine.

Jobs had no use for small-minded naysayers. His experience had taught him that if you offered a better computer, well priced and accessible, there was no limit to what human ingenuity could achieve with it. No one, after all, had thought of electronic spreadsheets when he and Wozniak rolled out the Apple II in 1977, but within two years, a spreadsheet program called VisiCalc—created in an attic by a first-year Harvard MBA student and a programmer friend—was one of the strongest drivers of Apple II sales. The PIC was not a consumer product like the Apple II, but the principle was the same. "People are inherently creative," Jobs remarked to an interviewer a few years later. "They will use tools in ways the toolmakers never thought possible."

Jobs directed Catmull and Smith to open sales offices for the Pixar Image Computer across the United States.

Pixar's small animation department—consisting of Lasseter, plus the part-time supporting efforts of several graphics scientists—was never meant to generate any revenue as far as Jobs was concerned. Catmull and Smith justified its existence on the basis that more films at SIGGRAPH like *André and Wally B.* would promote the company's computers. The group had no film at SIGGRAPH the preceding year, its last year under Lucas's wing, apart from the stained-glass knight sequence from *Young Sherlock Holmes;* Catmull was determined that Pixar would have a film to show at its first SIGGRAPH as an independent company in August 1986. (Among the films shown at SIGGRAPH in 1985, Lasseter particularly admired a piece of character animation called *Tony de Peltrie,* from a group at the University of Montreal; it featured a strikingly expressive human character, an aging piano player who entertained while inwardly reflecting on better days.)

The film would come from Lasseter's experiments with modeling his Luxo lamp. He had felt inspiration strike when Tom Porter brought his infant son into work one day and Lasseter, playing with the child, became fascinated by his proportions. A baby's head was huge compared with the rest of its body, Lasseter realized. It struck Lasseter's funny bone and he began to wonder what a young *lamp* would look like. He fiddled with the dimensions of all the parts of his Luxo model—all but the bulb, since lightbulbs come from a store and don't grow, he reasoned—and he emerged with a second character, Luxo Jr.

Lasseter initially intended the film as a plotless character study. When he showed some early tests at an animation festival in Brussels, a respected Belgian animator named Raoul Servais exhorted him, "No matter how short it is, it should have a beginning, a middle, and an end. Don't forget the story." Lasseter protested that the film would be too short for a story. "You can tell a story in ten seconds," Servais responded.

Lasseter was convinced. He devised a simple plot line in which the two lamps would play a game of catch with an inflated ball; Luxo Jr. would then approach the ball, hop onto it, bounce until the ball popped under him, and show dejection as the parent lamp looked on. Finally, Luxo Jr. would reappear feeling excited with a new, larger ball. Apart from the film's hoped-for promotional value, Catmull and Smith rationalized the project as a test of "self-shadowing" in the rendering software—that is, the ability of objects to shed light and shadows on themselves.

Because time and money were tight, Lasseter reduced the setting to its simplest elements. The background would be plain black. There would be no fancy camera moves to be worked out (in fact, no camera movement at all). His energies were focused instead on working out techniques based on classic animation principles to convey emotion. Even though the characters were faceless and word-

John Lasseter, at Lucasfilm in 1985, animating his Luxo lamp as an exercise. The right-hand video monitor shows a wire-frame model of the lamp. The pen and tablet he is using are digital, similar in function to a computer mouse. The Luxo project would eventually become the short film *Luxo Jr.*
© Roger Ressmeyer/CORBIS

less, Lasseter shaped such subtleties as the speed of the child's hops and the way it carried its head to convey in an instant when the child was feeling joy and when it was feeling sadness. At every moment, the parent and child each seemed to have a definite frame of mind.

While Lasseter worked on *Luxo Jr.,* two of Pixar's graphics experts were working on shorter films of their own with Lasseter's feedback. Bill Reeves, who was interested in algorithms to re-create the turbulence of ocean waves, made *Flags and Waves,* with waves reflecting a sunset and lapping against the shore. Eben Ostby, a Brown architecture major turned graphics programmer, made *Beach Chair,* starring a chair that walked across the sand and nervously approached the water, dipped its front legs in just far enough to test the temperature, and then scurried along. Reeves and Ostby also assisted Lasseter with model making and rendering on *Luxo Jr.*

All three films premiered at SIGGRAPH in the Dallas Convention Center Arena, where the audience of six thousand immediately recognized *Luxo Jr.* as a breakthrough. It had a far more realistic look than *André and Wally B.* (Pixar's marketing department did not go out of its way to point out that none of the film, not a single frame, had been rendered on a Pixar Image Computer.) More significant than its photorealism, however, was its emotional realism. It was perhaps the first computer-animated film that enabled viewers to forget they were watching computer animation.

Afterward, Lasseter saw Jim Blinn approaching him, obviously getting ready to ask a question. (Blinn was now back at the Jet Propulsion Laboratory.) Lasseter braced for a question about the shadowing algorithm or some other recondite technical issue that he knew equally little about.

"John," Blinn asked, "was the big lamp the mother or the father?"

It was true proof, Lasseter realized, that he had succeeded in applying the Disney touch of thought and emotion to his characters. (No one remembers how Lasseter answered Blinn's question, but Lasseter has elsewhere referred to the parent lamp as "Dad.")

In one respect, the twenty-nine-year-old animator had surpassed Uncle Walt. Disney had observed that giving lifelike qualities to inanimate objects held *comic* potential; "portrayal of human sensations by inanimate objects such as steam shovels and rocking chairs never fails to provoke laughter," he once remarked. Lasseter had sought this purely comic value in his student film *Lady and the Lamp.* On display in *Luxo Jr.* was a further insight, Lasseter's stroke of genius: that inanimate objects as characters held the potential for *dramatic* value. If the animator understood and applied the animation principles of Disney's Nine Old Men, objects could engross audiences with their emotions; they could appear, indeed, more human than humans.

Another important achievement for Pixar that year, a milestone that would prove crucial to its future, was the completion of the Computer Animation Production System (CAPS) deal with the recently renamed Walt Disney Company. The deal had been more than a year in the making. Once Disney became serious about replacing its ink-and-paint process with computers, it had reached out not only to Catmull and Smith at Lucasfilm, but also to their old rivals, John Whitney, Jr., and Gary Demos. Disney required the two teams to undergo a head-to-head competition. Smith's experience with Alex Schure's cel animators at NYIT turned out to be buried treasure from his past; he wrote a lengthy, detailed proposal that clearly conveyed that his group understood not only computer graphics, but also, more important, the production processes of traditional animation. He got the nod, and with Lucasfilm's approval, Pixar took the business with it when it spun out.

The negotiations that followed were difficult, not only because Disney under Eisner was aggressive in negotiations generally, but also because the project was taking Disney into something it had never done before. In the end, after the contract was signed, Roy Disney celebrated with Catmull and Smith over dinner in a private dining room hidden in the New Orleans section of Disneyland.

Smith needed someone to run the project and decided that the best candidate was a gifted engineer named Tom Hahn, an expert in image processing. Hahn, like many top engineers in computer graphics before him, had already passed word that he was interested in working for Pixar. The hire was awkward because it meant recruiting Hahn away from Smith's old friend Dick Shoup, the man who had introduced Smith to computer graphics at Xerox PARC a dozen years earlier. Shoup, who had since left Xerox to start his own company, was upset to lose a star talent. Nonetheless, there was little he could do to beat an offer from Pixar, which still had the cachet of its former connection with Lucasfilm. "The lure of Lucasfilm-slash-Pixar was just too much for people to ignore," Smith recalled.

The resulting system, developed by Hahn and a small team under him, used Pixar Image Computers to scan pencil drawings of characters, color them, composite them onto scanned backgrounds and other image layers, and record the frames onto film. It had its first test in 1988 with a single scene of *The Little Mermaid* (1989)—the concluding scene in which King Triton waves good-bye to Ariel and her groom. Disney, delighted with the results, surprised Pixar by immediately switching all of its feature animation work to CAPS, starting with *The Rescuers Down Under* (1990). Disney insisted on secrecy, fearing that the public would perceive computers as diluting the quality of the handcrafted films.

The truth was just the opposite; CAPS allowed Disney to equal and even outdo the graphical richness of the films from its golden age in the 1930s and 1940s. Most notably, Disney filmmakers gained significant flexibility because they no longer had to set colors according to the cel layer. Traditional methods required that multiple layers of cels be stacked under the camera, each with different characters or different parts of the same character. (Body parts that were in motion typically went on the top layers so they could be swapped in and out easily.) The problem was that the blank cels were not perfectly transparent, so they added their own hues to the layers underneath; a given shade painted on one of the middle layers ended up looking different from the identical shade on the top or bottom layer.

Thus, each color in the film—the color of a character's skin, for instance—needed to be painted with a different shade for each layer on which it appeared. Whenever a limb or a character went from one layer to another, its shade of paint had to change with it. With upward of 400,000 cels going into a feature film, the need for layer-coded painting was a significant burden. Moreover, the animators were effectively limited to five layers or thereabouts, because a deeper stack of cels would tint the bottom layers too much for any correction.

CAPS did away with all these limitations, slashing the effort needed to color the frames (which became a digital process) and effectively giving artists an unlimited number of cel layers. CAPS also allowed for digital special effects, which brought once-costly shots within easy reach. Multiplane-style shots, with their illusion of depth, could be created entirely within the computer, no longer requiring the massive and labor-intensive multiplane camera. *The*

Little Mermaid had three multiplane shots, all that its budget could accommodate; *The Lion King,* five years later, had hundreds.

Many at Pixar felt frustrated by Disney's secrecy—prestige and fame, more than money, being the currency of the research world from which Pixar recruited. "We were screeching [at Disney] for recognition," Smith remembered. Disney relented several years later, and employees of the two companies shared a technical Academy Award for CAPS in 1991.

The immediate effect of the CAPS project was to prop up Pixar's hardware business; with dozens of Pixar Image Computers running the CAPS software, Disney was the top customer of Pixar's computers. In the longer term, CAPS was the seed of Pixar's working relationship with Disney; it had surpassed Disney's expectations and had come in ahead of schedule. It was the first in a series of successful pieces of work that would serve as calling cards for Pixar when it was ready, years later, to move up to feature films.

The deal was far from enough to make the company profitable, however. Sales of the PIC had been disappointing after the early burst of activity. The company began work in 1987 on a lower-cost machine, the Pixar Image Computer II, with less capacity for expansion. Jobs irked Catmull and Smith by insisting that the machine's case be designed by Hartmut Esslinger's firm frog design (lowercase *f,* lowercase *d*) at a six-figure fee. Jobs was an enthusiast of the firm, which had previously designed the Apple IIc and the black NeXT Cube. Catmull and Smith appreciated fine design, but felt the contract was an extravagance for a struggling company. In the end, the design relied mainly on Lasseter's dimpled-square figure anyway.

Pixar rolled out the machine, undeniably sleek, in early 1988 with a price of $29,500. The price of the original model, by this time, had dropped to $49,000. Sales continued to crawl. The company had created an incredible piece of technology, but the world, it seemed, simply wasn't interested.

While work on the new machine was underway in 1987, the engineer who was in charge of the Pixar Image Computer software—a biophysics Ph.D. named Pat Hanrahan—came to

The Pixar Image Computer's lower-cost successor, the Pixar Image Computer II (shown with monitor). The market greeted it, too, with indifference.

believe that the company's real prize asset was its rendering program, Reyes. The program was in its fourth version, and it had proven itself by rendering *André and Wally B.* and *Luxo Jr.* He began lobbying for the chance to turn Reyes into a product.

"We had totally unique algorithms in the software that nobody else had," Hanrahan recalled. "I always thought that was the crown jewel of the company. So my argument was, 'Of all the great ideas and people we have here, we're trying to sell something that really isn't our core expertise—we should try to sell something where we actually have technology already developed and it's *in* our core expertise.' "

Hanrahan had met some people from Adobe around that time and was impressed with its PostScript language for laser printers. It struck him that Pixar could create an interface like the PostScript language, but for 3-D rendering.

A new language for 3-D images would fit perfectly with another idea that had kicked around the group back at Lucasfilm, namely, a specialized computer for rendering in 3-D at high speed. People referred to this concept, far more ambitious than the Pixar Image Computer, as the Reyes Machine or the Pixar-3D. It was the kind of

An early advertisement for PhotoRealistic RenderMan, introduced in 1989.

machine you would want if you were making a computer-animated film. The Reyes program itself had originally been written as a way to work out the internal logic of 3-D rendering hardware.

As Hanrahan saw it, Pixar could combine his idea of a new language for 3-D graphics with that of the Reyes Machine. Pixar could build devices to take images in the new language for 3-D graphics and render them at high speed, the way the Apple LaserWriter printer rendered PostScript images. The notion took still another turn when he invited a virtual-reality pioneer, Jaron Lanier, to give a talk at Pixar, and the two brainstormed afterward. Hanrahan and Lanier came up with the idea of a small device that one could wear and carry around—similar to the Sony Discman, the portable CD player popular at the time. It would generate movie-quality 3-D images that the user could watch through virtual-reality goggles. Pixar couldn't actually build it, not in 1987, but the two men thought it could be a hit when the technology caught up with the idea in five or ten years. They named it for the Discman, calling it the RenderMan.

Hanrahan won approval for the idea of a 3-D graphics language, and he and Bill Reeves began putting it together. "A lot of what I did for the next six months was just talking with everyone I knew in computer graphics who was using rendering software, and asking how they would like to use such a system and what features they would need to have," Hanrahan recalled.

The language Hanrahan and Reeves created was a highly general and powerful way to describe the shapes of complex objects and their positions in a scene. A separate language within it allowed the user to write programs, called "shaders," to define the appearances of surfaces and how they affect the light that falls on them; the color, intensity, and orientation of light sources; and the look of atmospheric effects such as fog. "It was sort of an unusual way of doing rendering because you let people change everything with this language," Hanrahan said.

The idea was that the new language—Hanrahan and Reeves called it simply Rendering Interface—would be the lingua franca of 3-D graphics. Users would create 3-D scenes using modeling software from other companies, and then that software would send the finished scenes to Pixar's software to be rendered. The language would be an open standard; if another company wanted to compete with Pixar by selling rendering software that understood Rendering Interface, that would be fine.

Catmull decided that there would be two announcements: one of the language (to encourage other companies to work with it), and another of Pixar's rendering software that used the language. Shortly before the language was to be announced in the spring of 1988, Jobs decided the name Rendering Interface sounded too dry. Someone remembered Hanrahan and Lanier's far-out idea of a "RenderMan" gadget, and that name was deemed suitably cool. Pixar and nineteen hardware and software companies announced their support for the RenderMan language in May.

At the same time, Smith, with his expertise in 2-D imagery, was working on a language called IceMan for processing digitized photographs and other images and composing new images from them.

About a year and a half later, in the fall of 1989, Pixar shipped its Reyes renderer, now retooled to work with the RenderMan language, under the name RenderMan Developer's Toolkit. (It was later called PhotoRealistic RenderMan, or PRMan for short.) The three-thousand-dollar product initially ran on Sun and Silicon Graphics workstations, and Pixar soon followed up with a version for the high-end Intel microprocessors of the time, the 80386 and 80486.

As with the Pixar Image Computer, Jobs's optimism for the new product and its market soared high. In a written statement, he declared, "Rendering is extremely important now as we expect it to become a standard part of all computers in the next 12 to 24 months."

Three-dimensional rendering, as Jobs saw it, was soon to take its place alongside desktop publishing as an instrument of communication. Pam Kerwin, hired that year from a graphics software company on the East Coast to serve as Pixar's vice president of marketing, remembered his enthusiasm for the idea. "He thought that RenderMan was going to help everyday people make photorealistic images on their computers," Kerwin recalled. "He thought that RenderMan was going to be a 3-D version of PostScript and these 3-D pictures would be flying out of people's printers."

In announcing the Intel version of the software, the company asserted, "Photorealistic three-dimensional images will soon be an essential form of communicating information in product design and development, marketing, animation, consumer product selection, and business communications."

The world, however, was not clamoring for photorealistic rendering. PhotoRealistic RenderMan was a brilliant technical success and was well received by the computer animation community and by special-effects operations such as ILM, but it remained a niche product.

"It's hard to know when you have new technology what the size of the market is," Hanrahan said. "But that software ended up being mostly for movie studios, whereas we initially thought it would be widely used in CAD and things like that, which never turned out to be the case. I don't think we understood that business."

For some, the experience was an introduction to Jobs's famed "reality distortion field"—his gift for making people in his presence believe anything.

"We'd have to deprogram our troops after he had made a visit, because he has this charisma," Smith remembered. "When he starts talking, he just takes people's minds. He would start talking to the Pixar people and I would just see their judgmental faculties go away. They would just sit there and look at this guy with what I would describe as love in their eyes."

Even foreknowledge of the reality distortion field was no defense. Kerwin recalled monthly meetings with Jobs at his NeXT Computer office in Redwood City, joined by Catmull, Smith, and Kolstad. They knew 3-D rendering was not ready for ordinary consumers yet; it was still a struggle even for Pixar's Ph.D. graphics experts. It was unclear that consumers even wanted it. "We were making these 3-D pictures and we all knew how hard it was," Kerwin remembered.

> And yet when we would go down to a meeting with Steve, he would be so convinced that this had Everyman potential that he would talk tough to you. He would say how this was really like PostScript and that we can have it in every printer, we can follow the Adobe model. So while you are in the room with him, you're thinking, *Well, yeah, that's true.*

> You actually *believe* it when you are there with him because he convinces you in a way that some of the things that you know are actually reality are really just that you are being shortsighted, or you are not trying hard enough, or you're just missing something. You believe him because he is so powerful and so charismatic and so enthusiastic. But then when you get back to the real world, you realize, *I knew that wasn't going to work.*

Catmull, as president, had the most dealings with Jobs. As time went on, Catmull found that role wearing, especially when there was bad news. "Ed would have a meeting with him and Steve, in his inimitable fashion, would go nonlinear [become unstable, lose his temper]," Kolstad said. "Ed was actually getting kind of physically ill dealing with Steve." (Executives at NeXT received similar treatment; on one occasion, witnesses remembered Jobs "screaming wildly" at a manufacturing manager who told him that the NeXT computer's black magnesium cases would cost two hundred dollars apiece, not the twenty that Jobs wanted them to cost.)

Catmull, Smith, and Kolstad finally agreed among themselves that Kolstad should take Catmull's place on the firing line. They persuaded Jobs to make the personnel change; Jobs promoted Kolstad to president and chief executive officer on December 1, 1988. Catmull became chief technology officer and board chairman. Kolstad, more seasoned than the others in this regard, took to meeting with Jobs on a weekly basis at NeXT so Jobs would hear fewer surprises. When Kolstad had occasion to recommend a plan of action, he picked up the habit of presenting a lesser idea first. "It seemed like his job was to trash your first idea—so you'd present your best idea *second.*"

As others pondered the company's vexing issue of revenue, Lasseter followed up *Luxo Jr.* with a short that was even more strongly character-driven, *Red's Dream.* Its star, Red, was a forlorn unicycle in a bike shop, left in a corner with a 50% OFF tag dangling from his seat. One rainy night, with the shop closed, the unicycle dreams of appearing in a circus and capturing the love of an audience; it awakens back in the shop, feeling worse, it seems, for having had a glimpse at a better life.

"Everyone begged me to give it a happy ending," Lasseter remembered.

The film project came with two technical rationales. The bike shop scenes at the beginning and end were to demonstrate the rendering of highly complex imagery; with the bikes and their spokes and the shop fixtures, a typical frame of the scene had more than ten thousand geometric primitives, which in turn were made up of more than thirty million polygons. (The idea of a bike shop setting was inspired by Eben Ostby, a cycling enthusiast, who had been working on generating a complex still image of a bicycle shop.) The dream sequence was to be a demonstration of rendering with the Pixar Image Computer. An engineer named Tony Apodaca had converted Pixar's rendering software to run on the PIC, but it turned out that the machine's design left its processors without enough memory for a program as complex as Reyes, and so Apodaca was able to convert only a portion of Reyes's features. On account of those limitations, the dream sequence was cruder in its look than the rest of the film, and *Red's Dream* was both the first and last Pixar film to be made with the Pixar Image Computer.

Space at Pixar was growing tight; while *Red's Dream* was underway, the animation group—Lasseter, plus several "technical directors" who created models, shaders, and the like—worked out of a hallway. Toward the end of production, Lasseter worked and slept in the hallways for days on end. One night about two weeks before the deadline for SIGGRAPH, an engineer named Jeff Mock brought his camcorder around and shot an ersatz interview with the bleary-eyed director:

> LASSETER: This is a Pixar film; it is *Red's Dream.* It is a new, totally new concept in cartoons. It employs the use of computers.
>
> Q: In what way?
>
> LASSETER: Well, we use the computers to make the film

and then we use a thing called a laser scanner to shoot the film and a thing called a laboratory to develop the film. It is quite revolutionary and I must ask you one thing.

Q: Yes?

LASSETER: May I shave your legs?

Lasseter had just spent five days animating a sequence of three hundred frames—twelve and a half seconds of film.

Frank Thomas and Ollie Johnston, the Disney legends, visited Lasseter at Pixar after *Red's Dream* had just been finished and they watched a screening. Thomas, evidently freed of the doubts about computer animation that he had expressed in his 1984 essay, shook Lasseter's hand afterward and said meaningfully, "John, you did it."

The film premiered in late July 1987, at SIGGRAPH in Anaheim, and was received with the enthusiasm customary, by then, for Lasseter animation. Another film of character animation ran at the same conference, *Stanley and Stella in "Breaking the Ice,"* featuring a bird and a fish that fall in love. John Whitney, Jr., and Gary Demos—head-to-head against Pixar once again—made the film with Craig Reynolds of Symbolics. It was technically interesting in some ways, but unlike the Canadian short *Tony de Peltrie* in 1985, it left the Pixar group cold. Lasseter felt that the fish, Stella, lacked emotion; she was animated in the sense of moving, but not in the sense of living and feeling, the result of a failure to heed the lessons of Disney's Nine Old Men.

Notwithstanding Lasseter's achievements, some of the engineers working in the trenches on the company's products wondered whether it made sense to keep the animation group going at all. As they saw things, they were working their hindquarters off to make money for Pixar while Lasseter's group only spent it. Their passion was building computers and software—stretching their skills in C, in bit-slice microcode, in gate arrays—not entertainment. Eventually they discerned, to their chagrin, the reason why the cash-strapped company was supporting an animation group: The real priority of the co-founders, Catmull and Smith, was to make films.

"In hindsight, it's clear they wanted to make movies and were pretending to be a computer company," said Rocky Offner, who was then a programmer working on a Macintosh version of RenderMan. "I was there for the software business, which I thought was important to Pixar, but I was clearly mistaken."

Bruce Perens, chief software engineer for the Image Computer II, was displeased to realize he was dedicating himself to an "interim business," as he put it. "I came there to work on Image Computers, not to be part of an animation studio."

The puzzle came together for Perens after hearing one technical staffer after another wonder why David DiFrancesco was there. Why did a computer company need a laser film recording expert? "For years, no one understood what David did that was necessary for David's being there," Perens said. "But they were just saving David because they felt eventually they'd need film recorder technology."

The engineers were not alone in wondering about the value of Lasseter's short films. On repeated occasions in the late 1980s, Catmull barely dissuaded Jobs from shutting down the animation division. Jobs's doubts were understandable: Pixar was reliably losing money every year and he was supporting the company through a line of credit with his personal guarantee.

One of those occasions came after the release of *Red's Dream.* When Catmull told Jobs that he planned for Pixar to prepare another short film, a skeptical Jobs came to Lasseter's office to hear the story. (Lasseter had gotten out of the hallway.) With Catmull and the animation group in attendance, the storyboards pinned on his wall, Lasseter went through the drawings and acted out the shots—much as story men had done on the Disney lot for decades. The stakes here, however, were higher. "We knew that he wasn't just pitching for the film, he was pitching for the survival of the group," said Ralph Guggenheim, who was managing the animation unit.

Tin Toy had much the same inspiration as *Luxo Jr.,* namely, Lasseter's observations of a friend's baby. This time, he opted for a far more ambitious tack, attempting to mimic a human baby in its appearance, the herky-jerky movements of its arms, its fickle moods.

With this he blended his longtime love of toys; he kept a collection of vintage toys at home. The story was told from the viewpoint of a toy one-man band, known within the production team as Tinny, who finds the baby charming at first and then frightening.

The film was officially a test of the PhotoRealistic RenderMan software. As with *Luxo Jr.* and *Red's Dream,* it was also a chance for Lasseter to one-up his earlier efforts, taking his animation and storytelling to another level. The baby proved very difficult to model and animate; "it just became an incredible burden," remembered Flip Phillips, a new member of the team at the time. In early attempts at a model of the baby's head, he appeared to have the face of a middle-aged man. The final version of the baby (known to the team as Billy) had a much-improved face, but his skin had the look of plastic. When he moved, moreover, his body lacked the natural give of baby fat and his diaper had the solidity of cement—compromises made necessary by lack of time and the still-developing technology.

Lasseter and his technical directors slept under their desks at times to get *Tin Toy* finished before SIGGRAPH in August 1988, but to no avail. What the SIGGRAPH audience saw was only the first three-fifths or so of the film, ending at a cliffhanger moment with Tinny running into his box and watching in horror through the box's cellophane window as Billy advances toward him. (A sharp-eyed viewer could spot a framed photo of nondigital origin on the living room coffee table—a photo of a young John Lasseter winning the "best boy camper" award.) "Even though it wasn't complete, people were wowed by it," Guggenheim remembered.

At the same conference, Apple Computer showed a short film for which Lasseter's new wife, Nancy Tague, was artistic director and one of several animators. Tague, a 1986 Carnegie Mellon graduate with a bachelor's in computer science, worked as a computer graphics engineer for Apple; the two had met at a SIGGRAPH conference several years earlier. *Pencil Test* was a demonstration of color graphics on the Macintosh II computer, centering on a Macintosh II pencil icon that comes to life, jumps off the screen, and then finds it cannot get back when someone turns the computer off. A recent CalArts graduate named Andrew Stanton, a writer on the short-lived animated series *Mighty Mouse, the New Adventures,* helped Tague with the story. The film used no Pixar software—Jobs's relationship with his former company was still adversarial—but Lasseter gave advice on the animation.

Pixar did partner with another company in co-sponsoring the hit party of SIGGRAPH that year. As SIGGRAPH had grown, huge parties—usually held by exhibiting companies—had become a staple of the conferences. For its party, Pixar combined with Pacific Data Images, another small Bay Area firm, which specialized in logos for television networks, cable channels, and local stations as well as commercials. The leaders of the two companies were friends, as were many of their employees. Their event was pool-themed, held in a ballroom with beach decor, beach balls for the guests, and hosts and hostesses dressed as lifeguards; the party was nearly shut down as beach balls kept hitting the chandeliers.

Tin Toy went on to win the 1988 Academy Award for best animated short film, Pixar's first Oscar. With the award, *Tin Toy* went far to establish computer animation as a legitimate artistic medium outside SIGGRAPH and the animation-festival film circuit. A member of the Academy's board of governors, animator William Littlejohn, saw in *Tin Toy* a window into the potential of the young medium. "There is a realism that's rather astonishing," he told *The New York Times.* "It emulates photography, but with artistic staging."

Robert Winquist, head of the character animation program at CalArts, went farther, predicting that computer animation was "going to take over in a short time." He publicly advised animators, "Put down your pencil and your paintbrush and do it another way."

The Academy Award also caught the attention of Lasseter's former employer, which made him a series of lucrative offers to come back as a director. For Lasseter, who was feeling squeezed financially as a married man with a stepson in elementary school and thoughts of more children to come, the offers presented a quandary. "I thought, 'Where am I going with this?' " he recalled. "I had won an Oscar for

Tin Toy, but I could hardly afford to have a family, and Disney was dangling a huge amount of money in front of me."

Yet he had powerful reasons to stay. He felt reluctant to uproot his new family and move to Los Angeles. He enjoyed exceptional creative freedom in his films at Pixar, and his brain trust for making computer-animated films was based there. Jeffrey Katzenberg, the head of Walt Disney Studios, had a reputation for being difficult and controlling. At Pixar, Lasseter had a comfortable chemistry with everyone he worked with, from Catmull and Smith to his technical directors and production coordinators.

"Everybody loved him," Pam Kerwin said. "He was respected. He had complete creative control. Yeah, he could have made lots more money at Disney, but these guys were motivated more by what they could do creatively than by how many dollars they made."

Lasseter turned Disney down. "We were aware they were trying to steal John away from us," Catmull said, "but John knew we had something important going on here. I remembered him saying, 'I can go to Disney and be a director, or I can stay here and make history.' "

The Oscar for *Tin Toy* also invigorated Jobs's interest in the animation group. He approved the production of another short. After the headaches of *Tin Toy,* Lasseter backed away from depicting human characters. "We all agreed, let's do something simpler that wouldn't drive us quite so crazy, and that could really be done in time, but that would be a lot of fun," Guggenheim remembered.

In a discussion with the group, Lasseter brought up the former MGM and Warner Bros. director Tex Avery, noting that his cartoons were wild and exuberant, yet not necessarily very complex. Lasseter collected snow globes and also enjoyed souvenirs from distant places; from those elements, *Knick Knack*—the only pure comedy among Lasseter's short films—began to fall into place. It told the story of a snowman in a snow globe who shares a shelf with curios from various hot spots of the world. He sees a bikini-clad figurine from Miami and becomes obsessed with breaking out of his plastic dome.

"There were a couple of sessions where we were all crammed into John's office," Phillips said. "He liked the idea of this guy trapped in the snow globe and his struggle to get out. We were all Chuck Jones and *Tom and Jerry* violent cartoon fans—the majority of us were, anyway—so the idea of escalating violence to get out came at these meetings."

Phillips and production coordinator Deirdre Warin simultaneously hit on the idea of the snow globe falling into a fishbowl, Phillips remembered, and Craig Good came up with the idea of an "iris out," a shrinking circle at the close, à la Looney Tunes.

A skeleton on the shelf held a surfboard and bore the inscription SURF DEATH VALLEY; the skeleton's 3-D model came from an Ohio State University skeleton data set called George, though the Pixar team stretched George's arms for comic effect. Also distorted were the two female characters—the bikini-attired woman and a mermaid—whose breasts were ultra-exaggerated thanks to a technical director who was a pinup enthusiast. (In Pixar's contemporary re-release of the film, the women's figures are of ordinary size.) In the same vein, a female toy under the sofa in *Tin Toy* was modeled with controls for her brassiere called "lift" and "separate," so that one could animate her breasts if one were so inclined.

The singer Bobby McFerrin created the musical sound track, which he improvised while watching a rough cut of the film. As the rough cut ended, the placeholder credits read *blah-blah-blah-blah,* so he sang *blah-blah-blah-blah* and it remained in the film's score. McFerrin did the score for free out of a belief that the film was cool to be involved with. (Gary Rydstrom of Lucasfilm likewise donated the sound effects of all Lasseter's short films from *Luxo Jr.* onward.)

Knick Knack premiered at the 1989 SIGGRAPH in Boston. It was one of the last pieces of animation that Lasseter would animate personally during Pixar's years as an independent company. When Lasseter presented it at the London Film Festival two years later, *The Independent* of London called it "a four-minute masterpiece" and the *Guardian* hailed Lasseter as "probably the closest thing to God that has ever graced the electronic images community."

. . .

Shortly after Kolstad became CEO in late 1988, he sat down with the animation team, together with Catmull and Smith, in the team's work area for a brainstorming session. The question was, how does animation at Pixar start earning its keep? How does it start putting dollars into Steve's wallet instead of taking them out?

"We looked at it as a group and said, 'Here's how we can make money,' " Kolstad remembered, "keeping in mind that the goal of the company and, really, the bliss of the company was to create a movie. That was a common dream. So it was to say: How can we bring that common dream to fruition and make money? Most of us at Pixar, excluding Steve, were pretty convinced that we could do that."

The idea of creating television commercials quickly emerged; advertising agencies had already been approaching Pixar thanks to Lasseter's short films. Others in the group objected. Advertising isn't creative, they said. People on the outside would come up with scripts and storyboards and we would just have to execute their ideas. That isn't what we're here for. "Then we don't take those kinds of jobs," Kolstad broke in. "We take jobs where we *can* be creative."

Lasseter favored the idea. Guggenheim wrote a short strategy memo outlining a three-step plan: Pixar would make commercials to let the animation team become self-supporting, then try a thirty- or sixty-minute television special to gain experience with production on a larger scale, and then finally graduate to a full-length feature film. Catmull and Smith were pleased to have a clear path to features.

Pixar had no experience selling itself as a production company and contracting with advertising clients, however. In July 1989, as Lasseter was wrapping up work on *Knick Knack,* Pixar entered into a contract with a large production house in San Francisco, Colossal Pictures, that did a lot of business making commercials; Colossal's sales force would represent Pixar to the ad industry.

Pixar's first project, the same year, was a fifteen-second shot of water lilies dancing to *Lohengrin;* it was a short tag to a commercial that aired in Japan for a large printing company, Toppan Printing.

The company also created its first complete commercial that year, titled *Wake Up,* for Tropicana orange juice. Pixar's ability to create photorealistic versions of inanimate objects and turn them into expressive characters was novel to television audiences, and other advertisers quickly took notice. In 1990, the pace accelerated as Pixar made commercials for the California Lottery (*Dancing Cards*), LifeSavers (*Skateboard*), Listerine (*Boxer*), Pillsbury (*Plump*), Trident gum (*Quite a Package*), and Volkswagen (*La Nouvelle Polo*). Despite Lasseter's Academy Award, the California Lottery's agency believed he had too little experience to direct a television commercial; hence, Colossal loaned Pixar one of its directors to direct *Dancing Cards.*

Lasseter added two animators that year to cope with the workload, seeking artists whose backgrounds were in traditional cel animation rather than computer animation. Stanton was the first; although he had been involved with *Pencil Test,* he was hired largely on the strength of his student films, which were popular on the festival circuit. Then, at the student film show at CalArts, which the Pixar animation group attended each year, Lasseter and his technical directors were impressed with the work of a graduating student named Pete Docter, who became Pixar's third animator.

"It was like, 'Man, we've got to hire this guy,' " Phillips remembered. "He had the whole package as an animator. He had great drawing skills and he had really great storytelling skills and sensitivity and he had really great timing."

As planned, the group was selective in the projects it took on, mainly accepting projects that offered a chance to do character animation and storytelling with inanimate objects. For example, its Listerine spot that year, inspired by the film *Raging Bull,* featured a yellow Listerine bottle suited up for a boxing match with gingivitis in a black-and-white ring. Newspaper stories about the fight spun onto the screen in the style of a 1930s film. As the bell rang, the bottle sent a punch into the camera—representing the point of view of the gingivitis—and the camera staggered as if it were dazed. The spot ended with the bottle holding the championship belt in triumph.

Pixar's television commercial production reached its peak the following year with fifteen spots, but the company would continue making them through the mid-1990s. Apart from the revenue the work brought in—$1.3 million in 1990, and a little more than $2 million annually in succeeding years—Pixar gained exposure and credibility. The commercials also created the chance to bring new talent on board and a training ground to prepare the newcomers for grander projects. The effort served, too, as a trial run at scaling up Pixar's pipeline from a single five-minute short a year.

"In the back of everyone's mind, especially in John Lasseter's mind, was that we were going to make a feature film at some point," Phillips recalled. One benefit of the commercials, he said, was that "we could learn basically what kind of production infrastructure we needed to make a big movie. It's one thing to talk a good talk, but to actually implement all this, what was it going to take? How did all this stuff scale?"

Pixar's hardware business, meanwhile, continued to languish. The Pixar Image Computer had never made the jump from a few early adopters—customers who readily embraced leading-edge products—to the mainstream. Part of the problem, Kerwin remembered, was that the company was targeting widely disparate niche markets, which made it harder to market the machine and harder to support it with software that made the PIC the answer to a customer's problems. "The market was not one big market, it was a whole bunch of small markets," Kerwin said.

Pixar did create sophisticated software that allowed the computer to combine radiological images from a patient, such as CAT scans, into a three-dimensional view, making them easier to read and interpret. Some radiologists became enthusiastic converts to the technique, but it was slow to gain acceptance.

Jobs himself blunted Pixar's entry into the medical market by antagonizing its partner, Philips Electronics, Kolstad recalled. Jobs could be cantankerous. "We had an executive come from Philips and demonstrate, et cetera, and then he got up to talk about the relationship. Steve got impatient and called him names." Philips never ordered another machine.

Finally, while Pixar's computers had some unique capabilities for image processing, the handwriting was on the wall that general-purpose workstations would soon be able to match them, or at least come close enough.

The machine's prospects seemed bleak. Kolstad won Jobs's approval to dispose of the Pixar Image Computer—the computer Kolstad had been hired to build—and announced the sale of Pixar's hardware business on April 30, 1990, to Fremont, California–based Vicom Systems (since defunct). Pixar would now focus on commercials and software.

The possibility of focusing on anything was disrupted when Lucasfilm unexpectedly gave Pixar notice to clear out of its quarters. Pixar had been subleasing the buildings on Kerner Boulevard from Lucasfilm since the spin-off, but now ILM needed the space. The old Lucasfilm hands at Pixar regretted having to leave. They could screen Pixar's films at the ILM screening room next door; the film recording guys at Pixar could drop by and talk with their counterparts at ILM. Lucasfilm, however, was adamant. Pixar scrambled to find a building that it could move into in time—"I think we missed by three days having a sheriff lock the doors on us," Kolstad said—and relocated to an office park across the Bay in Point Richmond.

Point Richmond was mostly a quaint town, but Pixar's new venue was near the adjoining city of Richmond proper, which was decayed and rife with violent crime. Down the road on West Cutting Boulevard was a liquor store called B&K Liquors—which, it was said around Pixar, stood for "Bleed & Kill." In the other direction, past some railroad tracks, was a Chevron refinery prone to occasional explosions and toxic gas releases; these would trigger periodic "shelter in place" lockdowns until the danger had passed.

Once ensconced in Point Richmond, the company pursued Jobs's vision of 3-D rendering everywhere with a series of less costly software products sold at retail stores. MacRenderMan, the first, was the Macintosh II version of PRMan. Another product, Typestry, created 3-D type with realistic surface textures and other special effects. Pixar 128 was a collection of texture images, such as wood panels, stones, and fabrics, for graphic artists. A package called Showplace allowed users to create 3-D scenes from prebuilt models of objects, ranging from spheres and cubes to furniture. The company's hope was that these products would break into the mainstream by finding their way onto the Macintoshes of graphics designers and desktop publishers. Like previous Pixar offerings, however, each either was too hard for its intended audience to master or else solved a problem that users didn't think they had.

By the year's end, Jobs was running low on patience. Pixar had lost money every year since he bought it, five years in a row, adding to the debt that Jobs was personally guaranteeing; in 1990 alone, the company ran a net operating loss of more than $8.3 million. There had been conversations with Disney about a film based on *Tin Toy,* but it was far from clear that these would lead to anything. The company's top managers sat down every month to figure out which bills absolutely had to be paid, and met with Jobs in his office at NeXT, hat in hand, to ask for permission to draw on the line of credit. "If, for example, we knew we were going to get a payment from the commercials that had been done, but we didn't have it yet and we really needed to pay our bills—it was the small-business drill," Kerwin remembered.

Adding to the tension of the situation was a simmering conflict in personalities between Jobs and Smith. Jobs accepted argumentation, up to a point, from employees whom he respected. Smith, though, seemed willing to approach the borders of Jobs's tolerance more readily than the others were. Smith saw Catmull and himself as Jobs's partners and equals as much as they were his employees.

"I think Alvy was a lot less comfortable than the rest of us with listening to Steve and then going home to try to do the best he can," Kerwin said. "He was more into saying, 'Well, wait a minute, I know what I'm doing.' "

Smith remembered that Jobs "was always jumping on Ed and me for not coming through with the numbers, then he would write the check and pay for it anyhow. But he would really let us have it. We were absorbing lots of abuse."

In one of those meetings, Jobs and Smith finally spiraled out of control. As Jobs was criticizing the Pixar managers for failing to hit a delivery date on a project, Smith interrupted and said, "Steve, but *you* haven't delivered your board on time"—meaning, a board for the NeXT computer.

It was the sort of remark Jobs normally might have put up with, but it seemed Smith had crossed a line by joking about Jobs's computer. "He went completely nonlinear," Smith recalled. "He went crazy on me and started insulting my accent."

Jobs had homed in on a sensitive spot. Smith's native southwestern accent, which he had mostly suppressed since his days as an academic in New York City, sometimes reemerged in moments of stress. Jobs mocked it.

"So I went nonlinear, too, which I had never done before or since," Smith remembered. "We're screaming at each other, and our faces are about three inches apart."

There was an unspoken understanding around Jobs that the whiteboard in his office was part of his personal space—no one else was to write on it. As the confrontation went on, Smith defiantly marched past him and started writing on the whiteboard. "*You can't do that,*" Jobs interjected. When Smith continued writing, Jobs stormed out of the room.

To outward appearances, the conflict blew over, but the men's relationship would never be the same.

In early 1991, Jobs decided he would continue to put money into the struggling company, but at a price: He would take back the employees' stock shares. Like typical Silicon Valley companies, Pixar used stock to motivate employees and to encourage them to stay. Pixar's program was based on restricted stock grants rather than stock options; an employee could buy a certain number of shares of the company for a low price, ten cents a share, and his or her ownership of the shares would vest over a period of years. Jobs, however, owned enough of the company to shut it down—on paper—and transfer its shares to a new company, "New Pixar," wholly owned by Jobs. The language of the stock grants had given Jobs the right to rescind them anytime, which he now did, offering employees the choice between the original price of ten cents a share or nothing. The original Pixar employees from Lucasfilm, Catmull and Smith on down, also lost the shares they had received during the spin-off.

Pixar's hallways resounded with bitter humor likening the stock contracts to toilet paper. Ordinarily, an infusion of capital would have merely diluted the employees' shares, not wiped them out. Many of the employees felt swindled, and the incident became literally a textbook case of abuse by a Valley entrepreneur.*

Few employees outside Pixar's top few managers, however, had any idea how dire the company's financial situation had become. Jobs, now the sole owner, decided still another drastic step was necessary.

PIXAR CAN'T SEEM TO ANIMATE ITSELF

Oscar-winning firm lays off 30—including its president

THE SAN FRANCISCO CHRONICLE
MARCH 29, 1991

By Ken Siegmann, Chronicle Staff Writer

Steve Jobs' Pixar Ltd., which makes highly sophisticated computer-graphics and animation software, has laid off 30 of its 72 employees—including its president.

Richmond-based Pixar won an Academy Award for computer animation in 1989. But insiders say that the company hasn't been profitable since Jobs, the Apple Computer co-founder and NeXT Inc. chairman, acquired it from film producer George Lucas in 1986.

. . . .

Part of Pixar's problem has been that its software technology exceeds the capabilities of available hardware. "The technology is sound, but nobody's ready for it yet," said analyst Joan-Carole Brigham, of International Data Corp. in Framingham, Mass.

* The textbook was Alan Hyde, *Working in Silicon Valley: Economic and Legal Analysis of a High-Velocity Labor Market* (Armonk, N.Y.: M. E. Sharpe, 2003), pp. 194–95.

Jobs shut down most of Pixar's activities outside of commercials and RenderMan development. Jobs believed there was a future in PhotoRealistic RenderMan as a product for printers and desktop computers, like PostScript, and in Alvy Ray Smith's IceMan program for image processing. Commercials had become the one profitable part of the company; these would subsidize PhotoRealistic RenderMan and IceMan until the market was ready for them. Employees who were not working in one of those areas lost their jobs without notice or severance.

"Our goal is to make RenderMan and IceMan the system software of the 90's," Jobs told *The New York Times.* "I've stuck with it this far, and the payoff looks like it should be in the next 24 months."

Kolstad was among the casualties. Kerwin, who was let go herself and then quickly rehired, was upset by the treatment of the departing employees; she argued to Jobs that they at least deserved a couple of weeks' notice. Jobs said, Fine, we'll give them two weeks' notice retroactive as of two weeks ago.

Jobs met with the remaining employees soon after the layoffs and brought his reality distortion field with him. "You're seeing your friends packing their stuff up and pushing it out to their cars," Phillips remembered, "and yet somehow he had convinced you that that was the greatest possible thing that could happen."

Within the Silicon Valley community, the talk was not of the way Jobs had handled his former employees at Pixar, but of his having kept Pixar going *at all.* It seemed to make little sense from a business point of view. For all his bravado about RenderMan, his motivation was likely a matter of status as much as economics. After his rise and fall at Apple, the onus was on him either to create another success story or to leave his peers to conclude that the first one had been a quirk of fate.

"It wasn't really working," Smith said of Pixar's early years. "In fact, that's being kind of gentle. We should have failed. But it seemed to me that Steve just would not suffer a defeat. He couldn't sustain it."

6. MAKING IT FLY—1

Toy Story

The engineers whom Pixar hired to carry out the CAPS project—Peter Nye, Michael Shantzis, and team leader Tom Hahn—spent considerable time in the Disney animation building in Glendale integrating the new digital-paint system into Disney's animation process. Early in the summer of 1990, as production was underway on *The Rescuers Down Under,* a message filtered back from the CAPS team: Peter Schneider, president of Walt Disney Feature Animation, was potentially interested in making a feature film with Pixar. It was the message that Ed Catmull and Alvy Ray Smith had been waiting a decade and a half to receive. Catmull hurriedly got on the phone with Schneider.

Pixar was not the only organization doing character animation with computers, but Pixar had gradually made itself Disney's obvious choice. With Lasseter's short films and commercials, and the successful introduction of the CAPS system, Pixar had created a series of highly visible calling cards. It was the quality of this body of work, rather than old-boy connections or an isolated lucky break, that had made Pixar the front-runner once Disney became interested in the medium—much as Ed Catmull's research and student film had made him the logical choice to start NYIT's computer graphics lab, much as Loren Carpenter's *Vol Libre* had made him an inevitable hire at Lucasfilm amid a horde of job seekers.

Catmull, Smith, and Guggenheim went to Glendale to meet with Schneider and with Thomas Schumacher, producer of *The Rescuers Down Under.* The Pixar group found the atmosphere puzzling. "We got a rather cool sort of reception from them where we were told, 'We know how to do it better than you guys,' " Guggenheim, Pixar's head of animation, recalled. "It became clear that basically they were being told by Jeffrey [Katzenberg, Schneider's boss] that they had to do this; they didn't like the idea of engaging an outside animation studio."

Catmull, Smith, and Guggenheim struggled to explain to the Disney executives that they would welcome Disney's creative help on a film. "We had no ego about the fact that we were happy to learn from them," Guggenheim said.

Nonetheless, Schneider remained cool. Guggenheim remembered, "The reception we got was so odd that we thought, *Well, maybe they don't want to do this.*"

Later, they learned the nub of the problem: Katzenberg intended that if Disney were to make a film with Pixar, it would be made entirely outside of Schneider's purview. Instead, the production would be an arm's-length relationship in which Katzenberg would simply agree with Pixar on a script and a budget and then Pixar would deliver a negative for the finished film when it was done; Disney Feature Animation would have no creative role. Part of the rationale was that Pixar was a non-union company, and it was unclear whether Disney's union contract would allow Disney to take part in making a non-union film. Schneider, in any case, was peeved.

After that first meeting, the Pixar contingent went home with low expectations, only to be surprised again when Katzenberg himself summoned them for another conference. This time, Catmull, Smith, and Guggenheim were joined by Bill Reeves (as head of animation research and development), Jobs, and Lasseter. They brought with them an idea for a half-hour television special called *A Tin Toy*

TOY STORY (1995)

The first fully computer-animated feature film was a highly uncertain gamble. Several major studios passed on the opportunity to make a film with Pixar before Disney took the chance. Even with Disney on board, Pixar owner Steve Jobs—nervous over cost overruns—was looking for investors to take the company off his hands as late as the year before the film's release.

As it turned out, *Toy Story* would be the highest-grossing film of the year and a critical favorite. Its script infused humanity into its toy stars and moved deftly from humor to pathos to adventure. In 2005, the National Film Preservation Board selected *Toy Story* for preservation by the Library of Congress in recognition of its historical significance.

A BUG'S LIFE (1998)

Inspired by Aesop's fable of the grasshopper and the ant, Pixar's second feature went head-to-head against DreamWorks's similarly themed *Antz* and won handily at the box office. The writers for *A Bug's Life* struggled with telling the stories of its numerous central characters—Flik, the leading ant; the ant queen and princesses; the circus bugs; and Hopper, the villain—within the confines of a ninety-minute feature.

TOY STORY 2 (1999)

The production of the sequel—originally a direct-to-video offering—brought hardships and near-tragedy to members of the crew. The results, however, impressed audiences; *Toy Story 2* was the rare sequel that seemed as original as the original.

The film explored new territory by delving into Woody's history, in which he had been a celebrity and part of a set of toys. The latter idea dated back to *A Tin Toy Christmas*, Pixar's planned (but never produced) television special of the early 1990s.

MONSTERS, INC. (2001)

Pixar's fourth feature envisioned a world in which monsters rely on scaring children to generat energy, but at the same time, the monsters regard everything from the human domain as toxic- to be kept out of monster territory at all costs. The company's graphics researchers developed self-shadowing hair for the character of Sully, achieving a new level of (stylized) realism. Pixar itself faced a different kind of monster in the form of two copyright infringement lawsui one of them filed by the cartoon artist Stanley Miller. The day before the film's long-announce release date, it was unclear whether the courts would permit *Monsters, Inc.* to open as planned

Christmas. They believed a television program would be a sensible way to gain experience with a project larger than a thirty-second commercial or a short film, and would give them the chance to build up their technical infrastructure. Once they had made a half-hour show, they reasoned, they could tackle a feature film.

As Catmull and Smith saw it, they had two hard challenges ahead at the meeting. One was keeping Katzenberg interested in working with Pixar. The other, perhaps more difficult, was selling Lasseter and the junior animators on the idea of working with Disney. Katzenberg had revitalized Disney animation after Michael Eisner put him in power in 1984, starting with *The Great Mouse Detective* and *The Little Mermaid,* but he had also built a reputation as a micromanaging tyrant in the process.

"Disney had such a bad reputation for how they treated the animators that these guys all elected to come to us instead," Smith recalled. "Ed and I said, 'Oh, good, here's our chance to finally make a movie'—but it required working with Disney, the company these guys had rejected. So the big problem was to get past that objection."

They met with Katzenberg at a long conference table in the Team Disney building, the company's headquarters in Burbank. Katzenberg understood the situation: He had to convince Lasseter that it was a good idea to work with him.

He began on a note that was at once pugnacious and flattering. "If I had my way, we wouldn't be doing this," he announced. "It's clear that the talent here is John Lasseter. And John, since you won't come work for me, I'm going to make it work this way."

He then spoke convincingly of his lifelong love of animation and how deeply he felt about great animated films. He said that he wanted Pixar to make its own films with Disney, not scripts that Disney would force upon it. He dismissed the idea of a half-hour television special; if you're going to staff up to do a holiday special and build a production infrastructure for it, he said, you might as well do a feature film.

Finally, he turned to the problem at hand. "Everybody thinks I'm a tyrant," he said. "I *am* a tyrant. But I'm usually right. I will give you a chance, if you think I'm wrong, to change the direction, but *you'd* better be right."

He invited the visitors to go mingle with animators at Disney and talk with them, right then and there. Ask them anything at all.

"We went off into rooms with the directors and the animators and other members of the movie animation team at Disney without Katzenberg in the room," Smith said.

> We just talked and talked and talked. And basically all of the people backed up what Jeffrey said, that you better believe he's a tyrant, there's no question about it, but his instincts are usually right on. They had seen instances of somebody sticking to their guns and changing the direction and it worked—but if it *didn't* work, their heads would roll.

The session came to a close and the six men walked out of the building, with Smith and Lasseter ahead of the rest. Smith, excited, asked Lasseter how he felt. "I think I can do it," Lasseter answered—meaning, he could work with Disney.

The two companies began negotiations, which soon turned into a logjam that continued for months. They were far apart on the division of revenues from the film, and Katzenberg also demanded that Pixar share the rights to its in-house software for 3-D modeling and animation, known as Menv (for "Modeling Environment"). Menv was fine-tuned to the needs of character animation. Unlike Pixar's rendering software, PhotoRealistic RenderMan, it was unavailable on the open market. To exert pressure on Katzenberg and give Pixar the best posture in the negotiations, Lasseter and Guggenheim met with executives from Universal, Paramount, and Columbia; none of them, however, offered Pixar a deal.

While negotiations were underway, Smith left Pixar to start his own company, Altamira, with seed money from the graphics software giant Autodesk. Smith's company was to develop advanced image processing software for personal computers. Since his confrontation with Jobs at NeXT headquarters the preceding year, he had felt viscerally that he needed to get Jobs out of his life; the fact that Jobs had taken everyone's Pixar stock added to his feeling of distrust.

In his talks with Disney, Jobs was unwavering on the issue of Pixar's technology, and finally Katzenberg accepted that Jobs was not going to give Disney rights to use Menv. The men homed in on the financial terms. Katzenberg had been receptive to the idea of a movie starring two toys; in March 1991, as the final contract appeared to be within sight, Lasseter gave Katzenberg a treatment for a film to be called *Toy Story.*

The original treatment for *Toy Story,* drafted by Lasseter, Andrew Stanton, and Pete Docter, had little in common with the eventual finished film. It paired Tinny, the one-man band from *Tin Toy,* with a ventriloquist's dummy (known only as "the dummy") and sent them on a sprawling odyssey, one that was to take them from the back of a truck to an auction, a garbage truck, a yard sale, a couple's house, and finally a kindergarten playground. Yet the core idea of *Toy Story* was present from the first treatment onward: that toys deeply want children to play with them, and that this desire drives their hopes, fears, and actions. The film, as Lasseter envisioned it at this point, was to begin with Tinny awakening in his factory:

TOY STORY (WORKING TITLE)

Everyone has had the traumatic childhood experience of losing a toy. Our story takes the toy's point of view as he loses and tries to regain the single thing most important to him: to be played with by children. This is the reason for the existence of all toys. It is the emotional foundation of their existence.

Our story opens in a small factory where tin toys are manufactured. They sweep down a conveyor belt with all types of other toys of different sizes, shapes and colors. We single out a group of tin toy musicians, then we focus in on one, our star, Tinny.

His eyes are opening up and he's looking around, wide-eyed with excitement. The conveyor belt goes to the shipping area where he's boxed up and put into a big carton. As the box is folded up, we follow him into the dark.

Tinny arrives at a toy store, where, as the store opens, he and the other toys are eager to be taken home, "like puppies in a pet store." He is given to a little boy for his birthday and is soon taken along on a road trip to the Southwest. The family accidentally leaves Tinny behind at a gas station, and he meets the ventriloquist's dummy, who greets him by imitating a cowboy. Ultimately, they end up as class toys in a kindergarten classroom. "As the kids show up, they start playing with the toys and the air is singing with the toys' joy," the treatment concluded. "Tinny and friends have really found their heaven, and their happy ending."

Apart from the premise concerning the toys' motivations, the most important concept from the treatment that made its way into the actual film was that of a toy getting left behind at a gas station. The other elements that survived were bits and pieces: Tinny being given as a birthday present, Tinny climbing into a truck, toys releasing the latch on a truck (the garbage truck in the treatment, a moving van in the film). A Slinky caterpillar in the treatment evolved into a Slinky dog in the film. Like the film, the treatment featured a threatening pet dog from which the toys must escape—a drooling, "disgusting" dog, Lasseter et al. wrote, "the type you want to kick across the room." The sadistic Sid Phillips of the film, however, was not yet on the scene.

Disney and Pixar reached accord on contract terms in an agreement dated May 3, 1991, and signed in early July.* Disney had been

* By coincidence, the film *Terminator 2: Judgment Day* opened a week before the announcement of the Disney–Pixar deal, ushering in a new era of computer graphics in special effects. Pixar had sought the contract to create the computer-animated water creature of *The Abyss* (1989), losing out to ILM; otherwise, Pixar had remained aloof from creating effects for live-action films since spinning off from Lucasfilm. Pixar did have a hand in special effects in one respect: Its PhotoRealistic RenderMan software was the all but universal choice of special-effects houses making computerized effects during this era from *The Abyss* onward.

DISNEY, PIXAR PACT ON 3-D ANIMATION

THE HOLLYWOOD REPORTER

JULY 12, 1991

By Greg Ptacek

The Walt Disney Company will become the first studio to distribute computer 3-D animated features in an exclusive three-picture deal signed on Thursday with Pixar, the company founded by Apple wunderkind Steve Jobs.

Pixar's Oscar-winning resident director-animator John Lasseter will write and direct the first feature, which will be released in 1994 under the Walt Disney Pictures banner and distributed by Buena Vista Pictures Distribution.

"Working with Disney to make the first full-length, computer-animated film has been a dream of ours since we founded the company in 1986," said Jobs. "Now our dream is realized and we couldn't be more excited."

hard-bottomed in the negotiations, in its usual style; the financial terms of the thirteen-page contract were lopsided enough that unless the film were a hit on the level of *The Little Mermaid,* Pixar's earnings from it would be insignificant. Disney retained the right "at its sole discretion" to "abandon the Picture at any time," even after production had already started. In that case, Pixar would receive for its labors only its incurred costs and a $350,000 "abandonment fee." While the contract was nominally a three-film deal, the second and third films were at Disney's option.

The contract gave Katzenberg final control over all creative decisions. If dissatisfied with Pixar's script, Disney could bring in screenwriters of its choosing. Disney would hold "100% ownership" of the film, as well as "sole discretion" over the making of sequels, remakes, television shows, and direct-to-video productions. Pixar would be first in line to serve as the production house for such projects, but if the two companies could not agree on terms, then Disney had the right to proceed on its own, or even with another production company such as Pacific Data Images; in that case, Pixar would get a minimal share of the revenues and it would have no creative say whatsoever.

Pixar kept making commercials as if the contract never happened—partly to keep the production staff gainfully occupied while waiting for the story to take shape, but mostly as insurance in case the film project went sour. "We recognized, as Disney did, that this was still a very experimental medium," said Guggenheim, who had become *Toy Story*'s producer. "There was no guarantee that Disney would stick through the whole process; they could bail out at any time."

The worry was not without foundation. Shortly after the contract was signed, Katzenberg put the film under the wing of Disney Feature Animation. The Pixar team was pleased that the move would give them an open door to counsel from Disney's animation veterans. Schneider, however, continued to take a dim view of the project and would later go over Katzenberg's head to urge Eisner to cancel it.

Katzenberg, for his part, was supportive of *Toy Story,* but felt that the initial treatment was problematic. Both of the lead characters wanted the same things for the same reasons. He told Lasseter to reshape *Toy Story* as more of an odd-couple buddy picture in the mold of *48 Hrs.* and *The Defiant Ones.* Both films centered on two men thrown together by circumstance and forced to cooperate in spite of their hostility, eventually gaining one another's respect. Lasseter, Stanton, and Docter emerged in early September with a second treatment. The lead characters were still Tinny and the dummy, but the outline of the final film was now beginning to appear.

TOY STORY

A musical toy named Tinny is given as a gift to two children. Like other toys, Tinny is "born" as he is unwrapped—naïve and innocent. Until now, a ventriloquist's dummy has been the children's favorite toy. When the children play with the toys, a wonderful world of make-believe materializes, enveloping both the children and their toys. As they play with Tinny, the dummy becomes jealous. He and Tinny begin to compete for the chil-

> dren's attention. They start to bicker, each one looking out only for himself.
>
> The children's father announces that the family is moving to a new town. After a hard day of packing, the family decides to go out for a pizza. The children bring along their favorite toys. Due to their selfishness, Tinny and the dummy fall out of the car at a gas station and are left behind. They find themselves completely alone in a big, frightening world, blaming each other for their predicament.

The two toys come close to catching up with the family at the pizza parlor, but instead they end up in the hands of "a mean kid who tortures his toys." The boy throws the dummy to his vicious dog and sends Tinny aloft on a model rocket. Both of them survive and plot an escape; just as they believe they are home free, they confront a new crisis as they find their owners' family leaving, its moving van pulling away. Tinny is able to get onto the van, but the dog catches hold of the dummy before he can make it. Tinny jumps off to distract the dog. The dummy, in turn, secures the help of the family's other toys, and together with the children they rescue Tinny.

The essentials of the plot were in place, but Lasseter was unsatisfied with Tinny as the lead character. "As the story evolved it became clear that Tinny was too antiquated," Lasseter recalled. "So we started to analyze what a little boy would get these days that would make him so excited that he stopped playing with anything else."

Tinny changed at first into a G.I. Joe–style action figure, the toy Lasseter liked most as a boy. Then he became a space hero along the lines of Major Matt Mason, another 1960s toy. The new character's name was Lunar Larry for a while, then Tempus from Morph.

With the thought that a buddy picture is most compelling if its lead characters are opposites, Lasseter then heightened the contrast of old and new between the two toys by making the ventriloquist's dummy a cowboy figure. The dummy-cowboy was named Woody after Woody Strode, an African American character actor who appeared in John Ford and Sergio Leone westerns.

Woody and Tempus made their debut in a thirty-second test that Lasseter delivered to Disney in June 1992. Disney had wanted a sample of what *Toy Story* would look like. The scene took place on top of a chest of drawers in the bedroom of Andy, their owner; Woody got Tempus out of the way as a rival for Andy's affections by tricking him into falling behind the chest, where he became stuck.

Foremost in the minds of the Pixar team was impressing Disney. "John very wisely wanted to feature in that film a number of things that could not be done in conventional hand-drawn animation," Guggenheim said. "So it took place in a room that was sort of dimly lit with venetian blind shadows falling across the room, with a character [Woody] who had a plaid shirt on, which you could never do in hand-drawn animation."

The characters were still far from their final incarnations in either their appearances or their personalities. Woody was devious and, with the ventriloquist's-dummy lines of his mouth, a bit spooky looking. Tempus from Morph, in a lobster-red space suit, was pint-sized, and more comical than tough. He was partly the deluded spaceman of the film and yet also aware that he was a toy with an owner.

The Disney executives were enthusiastic about the test, but objected to Woody's identity as a dummy; ventriloquist's dummies had become a staple of horror stories in film and television.

Lasseter turned for fresh inspiration to another toy from his boyhood, a Casper the Friendly Ghost doll with a pull-string voice. Woody reemerged as a stuffed toy with a pull string—and no separated jaw. Tempus became closer to Woody's size, making him a more formidable antagonist, and took the name Buzz Lightyear, inspired by the astronaut Buzz Aldrin.

Lasseter and the others on Pixar's story team—Stanton, Docter, and Joe Ranft—were keenly aware that most of them were beginners at the craft of feature-film writing. Indeed, apart from Ranft, who had done storyboard work on several Disney features and on the 2-D production of *The Brave Little Toaster,* and who had taught the story class at CalArts, none of them had any feature story or writing credits to their names at all. There was not necessarily a straightforward path between crafting a brilliant five-minute short and plotting and writing an eighty- or ninety-minute feature. Moreover, none of Lasseter's shorts from *André and Wally B.* onward had contained a single line of dialogue.

Seeking insight, Lasseter and Docter attended a three-day seminar in Los Angeles given by screenwriting guru Robert McKee. They came back to Point Richmond as true believers in McKee's principles, grounded in Aristotle's *Poetics.* High among these was McKee's doctrine that a protagonist and his story become interesting only as much as the forces arrayed against him *make* him interesting; character emerges most realistically and compellingly from the choices that the protagonist makes in reaction to his problems. A McKee seminar resounded with the master's observations about story structure and how it related to the progression of the hero's problems and his responses to those problems. McKee's teachings became the law of the land at Pixar.

Lasseter and his story department also screened a series of buddy pictures, including not only *48 Hrs.* and *The Defiant Ones,* but also *Midnight Run* and *Thelma & Louise,* among others, to study their structure. A longtime enthusiast of the Japanese animator Hayao Miyazaki, Lasseter screened Miyazaki's 1986 adventure film *Castle in the Sky,* as well. (Much later, when Lasseter started planning the chase at the end of the film, he assembled a reel of the chases from *The French Connection, Bullitt, To Live and Die in L.A.,* and other films to observe the directors' choices.)

Disney exercised its right to install outside screenwriters, hiring the comedy-writing team of Joel Cohen and *National Lampoon* alumnus Alec Sokolow to work with Pixar's story team on the script. After seven drafts, Cohen and Sokolow left and Disney brought in another screenwriter—Joss Whedon, a third-generation television writer, who spent about four months rewriting. What Whedon found, he later recalled, was "a great structure with a script that doesn't work."

The toy characters were mostly in place by the time Whedon

arrived. Apart from Woody and Buzz, the toys in the room were mostly classics familiar to baby-boom parents, or variations on classics; Lasseter feared that faddish toys would soon give the film a dated, musty feel. Hasbro had refused to license G.I. Joe and only reluctantly licensed Mr. Potato Head. ("There was a running gag in the production team about the endless conversations with Mr. Potato Head's lawyers to get him to appear in our film," Guggenheim said.) Whedon added Rex, the diffident dinosaur, and sought a pivotal role for Barbie.

As Whedon pictured it, Woody and Buzz, seemingly doomed at Sid's house, would be rescued by Barbie in a commando-style raid. Her character was to be patterned after Linda Hamilton's portrayal of Sarah Connor in *Terminator 2*. (The concept of a two-fisted, derriere-kicking heroine, still a novelty at the time, had also featured in Whedon's script for the 1992 film version of *Buffy the Vampire Slayer*.) Whedon's vision came to naught, however, when Mattel refused to license Barbie.

In addition to the fitful development of the script, another issue that concerned Disney was Lasseter's position on the use of music. The latest Disney animated hits resembled Broadway musicals, with characters breaking into song, invariably including an "I want" song in which the protagonist revealed his or her yearnings. But Disney's reliance on songs went deeper than that; songs like "When You Wish Upon a Star" were intrinsic to the Disney film tradition, conveying emotions in a powerful way. Lasseter resisted the musical-theater approach that Disney wanted, believing that musical numbers would take *Toy Story* out of its reality. Disney and Pixar reached a compromise: The characters in *Toy Story* would not break into song, but the film would use songs over the action, as in *The Graduate*, to convey and amplify the emotions that Buzz and Woody were feeling.

After Katzenberg approved the script on January 19, 1993, Lasseter began casting voices. To gauge how an actor's voice would fit with a character, he borrowed a technique from Disney Feature Animation: Take a scene from one of the actor's films that emblemizes the character and an aspect of the performance needed for the animated film, and then mate the actor's vocal performance in that scene with some storyboard drawings or finished animation. The result was a melding of the actor's voice with the appearance and actions of the animated character.

For Woody, Lasseter wanted Tom Hanks—well known in 1993, mostly for comic roles, but not yet an Academy Award winner. Lasseter was impressed with his performance in *A League of Their Own* the previous year. "What I loved about Tom was his ability to make all kinds of emotions appealing," he said. "Even when he's yelling at somebody, he's likable. That was crucial because Woody behaves pretty badly when he's not head toy anymore."

To give Lasseter the chance to see a spectrum of emotions in Hanks's performance, the test footage used his voice from a scene in *Turner & Hooch,* in which his character starts out calm and then rapidly grows hysterical as he pleads with his dog, "Not the car. Don't eat the car. *Not the car.*"

Katzenberg loved the test and agreed that Hanks was right for the part. He arranged for a meeting with Hanks, Lasseter, and Guggenheim at Disney's animation building in Glendale. Lasseter and Guggenheim filled the room with artwork for the movie. They showed Hanks a videotape of the test and explained the role of Woody; he was excited and signed on. (Guggenheim remembered that when Hanks walked into the room, no one realized at first who he was; he had just finished filming *Philadelphia,* in which he played a man dying of AIDS, and he was so gaunt as to defy recognition.)

Lasseter hoped to engage Billy Crystal to play Buzz, and created a test using a clip of Crystal's voice from *When Harry Met Sally;* Lasseter animated the test himself. When Crystal turned the part down, Lasseter turned to another actor. Tim Allen, a former stand-up comic in Los Angeles, had come to Katzenberg's attention via a half-hour cable special called *Men Are Pigs;* Katzenberg had viewed him as a potential lead for a television comedy series. After Allen turned down pilots for two other shows, he signed up for the Disney-produced *Home Improvement,* which premiered in the fall of 1991.

Lasseter liked the misplaced cocky self-confidence of Allen's onscreen persona.

Allen took the role of Buzz and quickly reshaped Buzz's personality. As far back as Buzz's incarnation as Lunar Larry, Lasseter and his team had imagined the character as a melodramatic save-the-day hero with the manner of a Dudley Do-Right. In Allen's first recording session, he delivered a different Buzz, one who was less caricatured, less self-important, and more of an ordinary guy. Lasseter was sold. "Instead of making Buzz aware of being a superhero, we made him more like a really good well-trained cop," he said.

Hanks, Allen, and the other actors recorded dialogue that Lasseter's team then married with some "scratch" dialogue (temporary dialogue voiced by non-cast-members, usually the animators) and scratch music to create story reels—a rough draft of the film in which a sequence of non-animated, hand-drawn story sketches took the place of finished footage.

The reels were a crucial step. In a live-action film, it is usual to film a scene with multiple takes for master shots and close-ups, different angles, different approaches to a piece of dialogue, leaving the choices among them to the director and editor in the editing room as they assemble the scenes. In animation—hand-drawn or computer-animated—each frame is costly, so filmmakers don't want to film lots of choices and then edit them afterward; they want to edit beforehand by way of the storyboards and, ultimately, the story reel. From the standpoint of a studio executive judging a work in progress, moreover, story reels are the first chance to see the story on screen, running at film speed.

Schneider saw the story reels on Friday, November 19, 1993, with Lasseter, Guggenheim, and Bonnie Arnold, a live-action producer whom Disney had sent to join the team as coproducer. Schneider deemed the film unready. On that day, dubbed "Black Friday" at Point Richmond, Schneider told the three of them to shut down production immediately. Animation, only recently started, could not resume unless and until they submitted a new script and Disney approved it.

They went back and gave the news to the production crew of twenty-odd people, many of whom had left other jobs to work on the project. The crew would shift to television commercials while *Toy Story* was in the garage. Some on the software team used the break to spend more time stabilizing Menv. Lasseter kept morale up by remaining outwardly buoyant, radiating the belief that the problems behind the shutdown were well within reach.

"The production shutdown was actually a scary time" for Lasseter, Stanton, Docter, and Ranft, remembered BZ Petroff, a hire from Colossal Productions who was managing the story department at that point. "Disney was reading them the riot act. It was, 'Get your story together *or else.*' But those guys handled it with confidence and laughter." They powered through the crisis, she said, relying on reserves of determination.

The critical problem in the story, they realized, was that they had made Woody unlikable: a mean, self-centered, sarcastic jerk. "He had to wind up selfless in the end, so our strategy had been, let's make him self*ish* in the beginning," Stanton said.

It was a reasonable strategy from a purely structural point of view—but it had led to a story that hardly anyone would have wanted to watch. The script had Woody abusing Slinky Dog, yelling at him, "Who said your job was to think, spring-weiner? If it wasn't for me, Andy wouldn't pay attention to you at all." Woody was to push Buzz out of the bedroom window intentionally, thereby sending him into Sid's yard, then close the blinds and remark to no one in particular, "Hey, it's a toy-eat-toy world."

Stanton retreated into a small, dark windowless office, emerging periodically with new script pages. He and the other story artists would then draw the shots on storyboards, and he would return to the isolation chamber to write more pages. Whedon also came back to Pixar for part of the shutdown to help with revising. By February 1994, Lasseter and his team believed they were on the way, and Katzenberg gave approval for Pixar to start production again in April.

The new script had several changes to make Woody more sympa-

Eben Ostby demonstrates digitizing a maquette of a character—here, Sid Phillips's dog, Scud. The Dalmatian belonged to an artist in the art department.
Photo by Louie Psihoyos/Science Faction

thetic. The film would open with a new sequence in which Andy played with Woody, highlighting the attachment between them. Where the old script showed Woody as a sneering overlord of Andy's room, early scenes in the new version would establish Woody as a sage leader who looks out for the other toys. The song "Strange Things," one of the three Randy Newman songs commissioned for the film, would play over a sequence revealing Woody's point of view as Buzz becomes Andy's new favorite. The film would ratchet down Woody's level of venality in the scene where Buzz falls out the window. Woody would no longer push him out; he would fall—hit by a swinging Luxo lamp, in a nod to *Luxo Jr.*—as the result of a lesser trick by Woody that had unforeseeably escalated out of control.

Lasseter showed the new story reels to the production crew.

"It was the scariest movie I'd ever seen," remembered Ronen

Barzel, one of the technical directors who would have to make it all happen in RenderMan.

With production having received a green light, the crew would quickly grow from its original size of 24 to 110, including 27 animators, 22 technical directors, and 61 other artists and engineers. Recruiting was brisk; the magnet for talent was not the pay, generally mediocre, but rather the allure of taking part in the first fully computer-animated feature film. "Disney gave us a very modest budget [$17.5 million] for *Toy Story,*" Guggenheim said. "Although that budget went up progressively over time, it didn't afford for very high salaries, unfortunately. We tried to make the other working conditions better. Just the enthusiasm of being able to work on a great project is as often as not what attracts artists and animators."

Most of the crew members were not actually employed by Pixar at all, but instead by a Pixar–Disney joint venture called Hi Tech Toons, formed to shield the two companies from liability and to simplify production accounting—a standard Hollywood practice. In the case of *Toy Story,* the separate company also took care of the union issue for Disney, enabling the studio to produce a non-union film through an entity that was nominally, at least, at arm's length.

As in traditional Disney animation, the organization of the work was a creative assembly line, though the steps in the processes were naturally different. Every character, every set, and every prop in *Toy Story* had to be designed by the art department, under Ralph Eggleston, and then turned into a 3-D model by one of the fifteen technical directors on the modeling team; they built more than four hundred models in all. The art department gave the modeling team pages with detailed drawings and written instructions, known as model packets. In creating the humans and the more complex toys, the modelers also digitized clay sculptures.

For any model that needed to be able to change its shape—such as a character with movable limbs and facial expressions, or an inanimate object that could be bent or scrunched—the model maker needed to rig the model with animation controls called articulated variables, or "avars." Avars represented places on the model that an animator could move. Buzz's spaceship-box needed avars so Buzz could open its door. Woody and Buzz had more than seven hundred avars apiece, including fifty-eight in Woody's expressive mouth. (Bill Reeves created the model for Woody; Eben Ostby modeled Buzz.)

Every shot in production passed through the hands of eight different teams. The art department gave a shot its color scheme and general lighting. The layout department, under Craig Good, then placed the models in the shot, framed the shot by setting the location of the virtual camera, and programmed any camera moves. As in Lasseter's short films, the layout team eschewed the elaborate, untethered flying moves often favored in computer graphics at the time. To make the new medium feel as familiar as possible, they sought to stay within the limits of what might be done in a live-action film with real cameras, tripods, dollies, and cranes. At times the group even emulated specific shots of live-action directors; a "Branagh-cam," as they called it, borrowed from Kenneth Branagh's 1994 *Frankenstein,* circling Woody at the moment when the other toys were concluding that he had pushed Buzz out the window. A "Michael Mann–cam," inspired by a technique used in the series *Miami Vice,* locked onto a wheel of a tanker truck as it pulled into a gas station and appeared ready to flatten Woody.

From layout, a shot went to the animation department, headed by directing animators Rich Quade and Ash Brannon. The work of the animators was conceptually the same on *Toy Story* as it had been on Pixar's shorts and commercials, breathing life into 3-D character models by applying the principles of Disney's Nine Old Men. The sheer magnitude of the effort in *Toy Story,* however, was vastly different from anything Pixar had done before, with seventy-seven minutes of animation spread across 1,561 shots.

Lasseter opted against Disney's approach of assigning an animator to work on a character throughout a film. At Pixar, by and large, animators were assigned to shots and animated all the characters in a given shot (normally three to seven seconds of film). Lasseter made a handful of exceptions in scenes where he felt acting was particularly critical, assigning animators to individual characters for all the shots in those scenes. One such scene was the confrontation between Woody and Buzz at the gas station; for that pivotal and highly emotional moment in the story, Lasseter assigned Woody to Mark Oftedal, a twenty-one-year-old who had been hired after his third year at CalArts, and Buzz to Guionne Leroy, an experienced Belgian-born animator who had come from a background in stop-motion.

In place of a pencil and drawing table, an animator used the Menv program to set the character into a desired pose. Once the animator created a sequence of hand-built poses, or "keyframes," the software would build the poses for the frames in between. The process was analogous to traditional animation, in which a senior artist would draw keyframes and then turn them over to a junior person whose lot was to draw in-between frames.

It was here at the animators' workstations, more than anyplace else, that the masses of binary data and meshes of polygons became conscious individuals with thoughts and emotions. The animators relied on their own acting instincts—"an animator is just an actor with a pencil," the old animation saying goes—and studied videotapes of the voice actors' line readings for inspiration. To develop the movement pattern of the green army men, whose plastic legs were attached to a base, an animator nailed a pair of sneakers to a board and practiced walking in them. Animators began with a character's broad body movements in a shot and only later dealt with facial expressions, thus forcing the animator to ponder how to convey the right emotions with body language alone.

The illusion of life in the characters was an accumulation of subtle choices by the animators. The eyes of computer-animated characters, especially, were the windows into their souls—no less than with live human beings—and required obsessive attention to detail.

"The eyes more than anything else give life to a toy," Lasseter said. "The angle of a blink, how far the pupils go off to the side when a character is trying to peek at something without being noticed, conveys a sense of presence better than any other element."

Lasseter rejected automatic lip-synching between the dialogue track and the character. Software could make a character's lips move in tandem with its words, but it could not make the lips fit the emotion of the moment; only an animator could do that.

The shading team, under Tom Porter, used RenderMan's shader language to create shader programs for each of a model's surfaces. Shaders, as noted, define a surface's color pattern, texture, and reflectivity; the team wrote some thirteen hundred shaders, ranging from a hardwood-floor shader to an Andy's-hair shader. Woody needed fifteen shaders to handle his face and his complex cowboy getup. Most shaders were self-contained, generating the look of the surface entirely through their own program logic—so-called "procedural" shading. Others, such as the shaders for Sid's paint- and scrape-strewn desk, relied on images that were painted electronically by the art department and then mapped onto the surface using the texture-mapping technique that Catmull had invented in graduate school. A few surfaces in *Toy Story* came from real objects; the shader for the curtain fabric in Andy's room, for example, used a scan of some actual cloth.

Ed Catmull with the "render farm" used to render *Toy Story.*
Photo by Louis Psihoyos/Science Faction

After animation and shading, the final lighting of a shot was in the hands of the lighting team, under Galyn Susman and Sharon Calahan. Lighting, by far the largest of the technical teams, had a lengthy set of objectives: to set the shot's mood, to indicate the time of day and perhaps the season, to guide the viewer's eye, and to underscore the characters' emotions and personalities. The lighting technical directors had a nearly infinite array of tools, including not only digital equivalents to all the lights available to a live-action film crew, but also lights far outside the reach of reality—lights that cast no shadows; lights that illuminate some objects in front of them, but not others; and lights that cast darkness instead of light, among other mind-bending, but highly useful, possibilities.

The completed shot then went into rendering on a "render farm" of 117 Sun Microsystems computers that ran twenty-four hours a day. Each frame took from forty-five minutes up to twenty hours to render, depending on its complexity (multiplied by more than 110,000 frames in the film). A camera team, aided by David DiFrancesco, recorded the frames onto film stock. Reflecting the long-standing belief at Pixar that factors like anti-aliasing made more of a difference to image quality than the use of thousands upon thousands of lines of resolution, *Toy Story* was rendered at a mere 1,536 by 922 pixels; on a typical cinema screen, each pixel corresponded to roughly a quarter inch by a quarter inch of screen area. To the human eye, the film would nonetheless look astonishingly crisp and clear.

The story team continued to touch up the script as production was underway. Among the late additions was the encounter between Buzz and the alien squeak toys at Pizza Planet, which emerged during a brainstorming session with a dozen directors, story artists, and animators from Disney. The script had Buzz drawn to a rocket-shaped structure that was actually a crane arcade game—but months of effort had not yielded any satisfying ideas about the toys he would find inside and what would happen there. (The rejected concepts had included toy bears and plastic pizza slices adorned with sunglasses.) As the meeting went on without resolution, someone uttered the key words: *the claw.* As if a spigot had been opened, ideas started flowing. "We began batting back and forth every mindless sect out of every movie we could think of," Stanton said. " 'Don't fight the claw,' 'do the will of the claw,' and so on." Chris Sanders, a story artist on *Beauty and the Beast,* sketched a distinctive three-eyed design for the aliens that carried over into the film.

Finished animation emerged in a steady drip of around three minutes a week. It was difficult for crew members to perceive the film's quality during much of the production process, when the finished footage was in scattered pieces and lacked elements like music and sound design.

"I had no reason to think it would be any good," recalled Barzel, who was then a recently minted California Institute of Technology Ph.D. on the lighting team. "I knew John was absolutely brilliant as an animator of shorts. But I've read authors who write good short stories and crummy novels; I figured it's a different skill. I had no reason to think John would have the skill to pull off a full-length movie."

He expected something that animators and animation buffs might find interesting, but that probably would not have a particularly wide audience.

"I joined because I wanted the practical experience," he said. "I thought, Well, it's going to be the first full-length [computer-animated] movie, so it'll be a fun thing to have been associated with, however it turns out."

What finally made Barzel a believer was watching Lasseter at work. He found that Lasseter had an uncanny ability to shift between the macro level of the entire film and the micro level of whatever detail he was dealing with at the moment. "Looking at an individual frame—it's meticulous work—he would always be aware of its role in the larger context of storytelling," Barzel recalled. "He'd say something like, 'This is the first time this character responds to that situation; it's really important that he get the right glint in his eye.' " Barzel started to think, *John knows what he's doing. This movie could be really good.*

One figure at Pixar, however, lacked that confidence in *Toy Story.*

7. MAKING IT FLY—2

Toy Story

Steve Jobs had put some $50 million into the company. It was still reliably losing money year after year. Now he also faced the possibility of millions more in liability; although Disney had agreed to increase its lowball $17.5 million budget for *Toy Story* to $21.1 million, it still wasn't enough. By 1994, costs were expected to run some $6 million higher. Hence, Disney forced Pixar to obtain a $3 million credit line to cover its share of the overages—backed, if necessary, by Jobs's personal guarantee. Weary of watching Pixar's deficits pile up, Jobs had tried to sell all or part of the company many times while *Toy Story* was in production: to Hallmark, to Microsoft co-founder Paul Allen, to Oracle CEO and co-founder Larry Ellison. Now, in the fall of 1994, Jobs looked to sell Pixar to another deep-pocketed prospect: the Redmond, Washington–based software colossus, Microsoft.

Microsoft was represented in the discussion by its senior vice president for advanced technology, thirty-five-year-old Nathan Myhrvold. After finishing his Ph.D. at Princeton at age twenty-three, Myhrvold had worked for a year as a postdoctoral fellow with the physicist Stephen Hawking at Cambridge, tackling theories of gravitation and curved space-time, before taking a three-month leave of absence to help some friends in the Bay Area with a software project. He became caught up in the excitement of personal computer software and entrepreneurship and never went back. In Berkeley, he co-founded a company called Dynamical Systems to develop operating systems for personal computers, which struggled for two years until Microsoft bought it in 1986. At Microsoft, he persuaded Bill Gates to let him establish a corporate research center, Microsoft Research, with Myhrvold himself in charge.

Myhrvold spent much of his time scouting for top researchers and promising new technology. He had met several of Pixar's technologists—Catmull, Smith, and Loren Carpenter—at SIGGRAPH conferences in the late 1980s and kept in touch.

"I was interested in them initially because we were interested in graphics, and we had the idea that maybe there's some technology that we could invest in early on that would be relevant to PCs later," Myhrvold recalled. "Then over time, as our interest got to be greater and Pixar was still sort of stumbling around trying to find its way, we had a number of discussions about whether Microsoft could license technology from them."

Thus, in the early 1990s, he had entered into talks with Pixar on several occasions about licensing PhotoRealistic RenderMan to be incorporated into Windows. As he envisioned it, PRMan would become the standard way for Windows programs to handle high-end graphics. Because PRMan was far too slow for interactive applications, such as games, the two companies hatched the idea of a fast subset of it for those applications, to be called Real-Time RenderMan. In the end, however, Myhrvold opted for another alternative; Microsoft licensed 3-D graphics software from Silicon Graphics called OpenGL—less powerful in some respects than Pixar's, but faster and more attuned to the needs of interactivity. (Later, confusingly, Microsoft would also give Windows still another 3-D graphics interface, known as Direct3D.)

Although the earlier conversations had not led to anything, Myhrvold was still keenly interested when Pixar came calling in

1994. Microsoft had just bought Alvy Ray Smith's company, Altamira, a few years after Smith left Pixar, and lured him to Microsoft Research as its first Graphics Fellow. Myhrvold was lining up other renowned computer graphics researchers. "The largest single cluster of incredibly great graphics people that we didn't have was Pixar," Myhrvold said.

Jobs and Gates, born the same year, had both been part of the personal computer industry since its beginnings. Jobs had gone out of his way in the past to insult Microsoft products in a way that seemed tinged with personal rivalry. ("Bill Gates has always had a huge respect for Steve Jobs," Myhrvold said. "At various points, it's been less clear if it's been reciprocal.") Now, however, Gates was not the enemy; Jobs's antagonism was fixed on the company from which he had been driven out, Apple Computer—an attitude exemplified by his office computer at Pixar, a laptop running Windows.

Myhrvold came to Pixar headquarters in Point Richmond to meet with Catmull and Pam Kerwin. Hoping to get him excited about *Toy Story,* they showed him one of the first completed scenes, the march of the green army men and their rope-drop from the second floor. He thought the scene was brilliant, and he was impressed with the cleverness of building a story around toys—plastic surfaces being ideally suited for computer animation. Still, he wasn't interested in Pixar for filmmaking; he wanted its technology and its graphics geniuses to bring Pixar-level graphics to Windows. On the other hand, Microsoft wouldn't *stop* them from making films, if that's what they wanted.

"I thought there would be a convergence between entertainment and PCs. You could extrapolate and say, 'If that's the future, then we'd like to have something like Pixar. And, yeah, they also want to make a movie, and maybe that'll work out.' "

Jobs, however, had a change of heart and decided not to sell after all. Instead, he merely licensed Microsoft several of Pixar's patents, for which Microsoft paid a onetime fee of $6.5 million; the patents covered, among other things, Pixar's techniques for anti-aliasing, motion blur, and realistic depth of field.

Kerwin, who was now vice president and general manager of Pixar's technology division, attributed Jobs's switch to a realization that something important was happening within Pixar's walls. "Steve kind of jerked back because, I think, there was something visceral in him that said, 'This is really going to be stupendous.' "

The consumer products arm of Disney—the group responsible for licensing toys and other tie-ins—was also slow to see the potential of *Toy Story.* It was a case of out of sight, out of mind: *Toy Story* was in production hundreds of miles away. Preoccupied with two other forthcoming releases, *Pocahontas* and *The Hunchback of Notre Dame,* Disney Consumer Products left the Pixar film on the back burner. When Guggenheim met with one of the division's senior licensing executives in December 1994, he was alarmed to discover that she saw no licensing potential in the film.

"We put together a presentation reel of scenes from the film that we'd already completed, and material on how the film was being made," Guggenheim said. "We were taking that around the company so people could get a feeling of what this film was all about."

The executive told him, I don't know how we're going to do toys for this.

"What do you mean?" Guggenheim queried. "It's *Toy Story.* You know, *Toy . . . Story.*"

Yes, she said, but you have all these toys that already exist—Mr. Potato Head, Speak & Spell, all that stuff. How are we ever going to make money off that?

"But you have all these original characters. You've got Buzz, you've got Woody."

She remained dubious. On top of that, both Disney and the toy companies had ended up with half as much lead time as usual. Schneider, now buoyant about the film, announced in January 1995 that *Toy Story* would release in the United States in November, coinciding with Thanksgiving weekend and the start of the winter holiday season. (Elsewhere, the film would open the following March.) Disney Consumer Products and the toy manufacturers were accustomed to eighteen months to two years. Mattel and Hasbro, believing that the runway was too short for them to have new toys ready in time, passed on the *Toy Story* license.

In February, Disney took the idea to Toy Fair, a toy industry trade show in New York. There, a Toronto-based company with a factory in China, Thinkway Toys, became interested. Although Thinkway was a small player in the industry, mainly producing toy banks in the form of film characters, it was able to scoop up the worldwide master license for *Toy Story* toys simply because no one else wanted it. The firm took the gamble and set to work on Woody and Buzz. Lasseter pressed Thinkway to produce the scary mutant toys of Sid's room as well, but the company demurred.

Jobs, meanwhile, was warming to *Toy Story* to such an extent that he had embraced a seemingly ludicrous idea: He was going to take Pixar public soon after *Toy Story*'s release. One financial adviser after another told him to forget about it. At the time, the notion of a public stock offering for a company that had never even turned a profit was alien to the thinking of serious investors.

Jobs would not be swayed. Jobs's confidence in Pixar's fortunes had grown when he attended a Disney press briefing in January. The briefing was held under a large tent in Central Park with a hundred-seat theater inside, where Mayor Rudy Giuliani and Michael Eisner announced that Disney's *Pocahontas* would have a public premiere in New York that summer. Lasseter and Guggenheim gave a brief presentation on *Toy Story* afterward.

That Disney had been able to stage the briefing in the middle of Central Park was impressive enough. What Jobs heard about the *Pocahontas* premiere impressed him even more; Disney had persuaded the city to let it take over the Great Lawn in Central Park in early June, fence it off, and put up seven-story-high cinema screens. Up to a hundred thousand people would be invited. The panache and extravagance of the event seemed amazing. Jobs had an immense flare for showmanship himself; his keynote conference speeches for NeXT and, earlier, Apple were always eagerly awaited for their slick theatricality as much as their content. He could respect what Disney was pulling off. He came away from the briefing having absorbed a lesson: When Disney gets behind something, *look out.*

Having heard enough in New York to conclude that Disney could turn *Toy Story* into a hit, he was determined to build up a war chest. With the proceeds of a successful IPO, he decided, Pixar would no longer have to accept the status of a mere contractor to Disney, supplicating for money, working for hire; instead, it could co-finance its subsequent films as an equal partner.

His first move, in February, was to strip Catmull of his title as president, the post he had held continuously since the company's founding nineteen years before (except for the two and a half years when Chuck Kolstad had stepped into his shoes). Jobs reasoned that an unknown figure like Catmull would hardly project an appealing image to investors. Jobs instead created an "Office of the President," a triumvirate that would consist of himself as chairman and chief executive officer; Catmull as senior vice president and chief technical officer; and a new hire, Lawrence Levy, a well-respected chief financial officer whom Jobs brought on board specifically to make Pixar ready to go public. Levy was to supply the credibility on Wall Street that Jobs lacked.

"At that point, people thought [Jobs] was a one-shot deal," Kerwin said. Conventional wisdom held that Jobs's success at Apple was a fluke, as evidenced by his two subsequent companies, Pixar and NeXT. "So Steve wasn't the guy to represent us to Wall Street, Lawrence was. Lawrence was just the right guy for the job because he could deal with Steve and he was very, very smart and he was very, very good at managing the expectations of institutional investors."

Catmull responded to his demotion without a trace of the petulance often exhibited by executives in such circumstances. His ambition appeared to be for the company more than for himself; he accepted that it was necessary to cater to Wall Street's perceptions. Meeting with a group of employees to announce the change, he joked that nobody wanted to be president and he hadn't been able to get anyone else to take the job. Not even Molly would do it, he said, referring to the company-mascot sheepdog who was in attendance.

Rounding out the executive ranks were Lasseter as vice president for creative development, Guggenheim as vice president for feature-film production, and Kerwin as vice president and general manager of the technology division.

As preparations for the public offering geared up in the spring, they would create a rupture in the company, opening a Pandora's box of gossip, resentment, and, in some cases, outrage. Until this time, there had been no riches to be had at Pixar and no particular expectation of any. Catmull and Lasseter were earning $160,000 and $140,000 a year, respectively—enough to put them in the comfortable upper middle class of the Bay Area, depending on one's definition, but modest for their positions by the standards of either Silicon Valley or Hollywood. Lasseter drove a well-worn blue Honda Civic.

The public offering process brought details of the intended allocation of Pixar stock options into view. A registration statement and other documents with financial data had to be prepared for the Securities and Exchange Commission and a prospectus needed to be made ready for potential investors. These documents had to be reviewed and edited, and it was here that the word apparently leaked: A small number of people were to receive low-cost options on enormous blocks of stock. Catmull, Levy, and Lasseter were to get options on 1.6 million shares apiece; Guggenheim and Reeves were to get 1 million and 840,000, respectively. If the company's shares sold at the then-planned price of fourteen dollars, the men would be instant multimillionaires.

The revelation was galling. Apart from the money, there was the symbolism: The options seemed to denigrate the years of work everyone else had put into the company. They gave a hollow feel to Pixar's labor-of-love camaraderie, its spirit that everyone was there to do cool work together. Also, it was hard not to notice that Levy, one of the top recipients, had just walked in the door.

"There was a big scene about all that because some people got huge amounts more than other people who had come at the same time period and who had made pretty significant contributions to the development of Pixar and the ability to make *Toy Story,*" Kerwin said. "People like Tom Porter and Eben Ostby and Loren Carpenter—guys that had been there since the beginning and were part of the brain trust."

Garden-variety employees would also get some options, but besides being far fewer, those options would vest over a four-year period. Even employees who had been with the organization since its Lucasfilm days a decade earlier—employees who had lost all their Pixar stock in the 1991 reorganization—would be starting their vesting clock at zero. In contrast, most of the options of Catmull, Lasseter, Guggenheim, and Reeves vested immediately—they could be turned into stock right away.

"I decided, 'Well, gee, I've been at this company eight years, and I'll have been here twelve years before I'm fully vested,' " one former employee remembered. " 'It doesn't sound like these guys are interested in my well-being.' A lot of this piled up and made me say, 'What am I doing? I'm sitting around here trying to make Steve Jobs richer in ways he doesn't even appreciate.' "

The big stock option grants had stemmed from Pixar's 1991 film deal with Disney. Because Disney made heavy use of employment contracts with its own employees, it demanded that Pixar put everybody on *Toy Story* on an employment contract as well. Pixar's leadership balked, believing that employment contracts would foul up the company's culture; Catmull's view had always been that employees should stay at Pixar because they wanted to, not because they were locked into a multiyear commitment. Disney relented considerably, but still insisted on contracts for those few whom it regarded as essential to the film. The 1991 Disney–Pixar contract thus required that Pixar enter into agreements with the so-called "key creative Pixar talent"—defined as Catmull, Lasseter, Guggenheim, and Reeves—to secure their services.

Each of the four men was now in a powerful position: no employment contract, no Disney deal. To prevail on them to sign, Jobs created a profit-sharing plan for them in February 1993. Sixteen percent of Pixar's profits on a film would go into a "profit pool," to be divided equally among the four. A couple of years later, as Jobs was planning for a public offering, they were at an advantage again; Jobs's advisers told him that investors would not accept such a rich profit-sharing plan for a handful of top employees, so he would need to induce them to give it up. On April 28, 1995, Jobs and the five executives agreed to replace the plan with munificent stock options. Although Levy had not been part of the profit pools, he too had the clout to insist on generous options.

The fact that there was a business rationale for the uneven treatment did little to mollify those outside the golden circle.

"Steve spent lots of time in people's offices, and I spent lots of time in people's offices, trying to explain that this is for the greater good of mankind," Kerwin said, "and if they win, we're all going to win. Some of them were threatening to leave."

Catmull, Kerwin, and Guggenheim drove to Palo Alto to get advice on the situation from the lawyer who was handling the public offering, Larry Sonsini.

When Sonsini graduated from the University of California at Berkeley's law school, Boalt Hall, in 1966, his classmates found his choice of jobs puzzling. While most of them were heading for large law firms and government agencies, he had taken the advice of a securities law professor and moved to Palo Alto, then regarded as a legal and business backwater. The professor, whom Sonsini had gotten to know as a research assistant, had alerted him to an entrepreneurial opportunity: Small technology firms were beginning to pop up in the South Bay and the nearby Santa Clara Valley. (The moniker Silicon Valley had not yet been coined.) The firms that prospered would need capital to grow. They would need securities lawyers.

Sonsini joined a three-lawyer firm where he was the first associate. In his early years of law practice, he spent half his time doing work for start-up companies and the rest of it on a typical small-town lawyer's diet of personal injury cases and the like. Building up the securities side of his practice proved an exercise in frustration; as his start-up clients grew and needed help with mergers or initial public offerings, they generally assumed they needed a big-city firm and he would lose them. He persevered over the next decade and gradually began bringing in IPO work. In 1978, Sonsini became chairman of the firm, which was renamed Wilson, Sonsini, Goodrich & Rosati. Two years later, Sonsini catapulted the firm into prominence by winning the role of counsel to a famous personal computer company in its initial public offering—Apple.

In 1995, the Wilson, Sonsini firm was far and away the dominant American law firm when it came to taking high-tech companies public, and Sonsini himself had become the lion of the Silicon Valley bar. His clientele had grown to include not only start-ups but also established companies such as Hewlett-Packard, investment banks, and venture capitalists. For Jobs, Sonsini was not just a lawyer, but a trusted business counselor.

Catmull, Kerwin, and Guggenheim laid out the problem in Sonsini's office. What do we tell the employees? they asked. How in the world do we manage this?

Sonsini told them calmly that they didn't have anything to worry about.

"Look, Steve is not going to take this company public," he explained. "He cannot take this company public. This company is fifty million dollars in deficit and has no revenue.

"He's not going to take this company public," he repeated.

On that subject, Jobs disregarded Sonsini's judgment. Sonsini had logic and experience on his side, while Jobs's idea was backed only by his own immovable will. Yet Jobs would turn out to be right. An event on August 9 would soon come to the rescue of Jobs's vision, namely, the initial public offering of Netscape Communications. The offering would introduce a topsy-turvy time in the history of the capital markets; investors who were well connected enough to buy shares of the unprofitable one-year-old company at the IPO price of twenty-eight dollars saw their investment more than doubled at closing time the first day. It would be all the precedent Jobs needed. Pixar, with all its losses, was a blue-chip enterprise compared to Netscape.

The edited *Toy Story* was due to Randy Newman and Gary Rydstrom by late September for their final work on the score and sound design, respectively. A test audience at a theater near Anaheim in late July indicated the need for last-minute tweaks, which added further pressure to the already frenetic final weeks. The audience reacted flatly to the film's opening scenes, leading Lasseter to resolve to punch the scenes up. Also, the film at that time ended with an exterior shot of Andy's house and the sound of a new puppy. Michael Eisner, who attended the screening, told Lasseter afterward that the film needed to end with a shot of Woody and Buzz together, reacting to the news of the puppy.

Pixar's future was now tied to a date, November 22, 1995, the day when *Toy Story* would release. It was two years almost to the day since Lasseter had shown the disastrous story reels at Disney. The story had been polished to a shine in the meanwhile, but it remained an open question how audiences would respond. While the response cards from the test audiences were encouraging, their ratings weren't at the top of the scale, either.

Disney's marketing engine was doing its part to get children in the theaters. Buena Vista Home Video put a trailer for the film on seven million copies of the video for *Cinderella;* the Disney Channel ran a television special on the making of *Toy Story;* Walt Disney World in Orlando held a daily *Toy Story* parade at Disney-MGM Studios. Also, while Disney Consumer Products had been slow out of the gate, Disney's film distribution arm, Buena Vista Pictures, was energetic in lining up comarketing agreements with promotional sponsors. Of the $145 million in promotional money for the film—more than five times the film's final production cost—only $20 million would come from Disney. The rest would be spent by Burger King, Nestlé, and other consumer products companies in return for the privilege of being associated with a Disney film. Archrivals

Pixar staff gathered in 1995, shortly before the release of *Toy Story,* in what was then the company's modest screening room. In the front row (L–R) are Craig Good, John Lasseter, Steve Jobs, and Ed Catmull.
Photo by Louie Psihoyos/Science Faction

PepsiCo and the Coca-Cola Co. both climbed aboard—PepsiCo through its Frito-Lay subsidiary and Coca-Cola through its Minute Maid unit. Starting a week before the release, Burger King offered *Toy Story* figurines and finger puppets for $1.99 with its Kids Meals.

Disney held an invitation-only premiere on Sunday, November 19, at the El Capitan, a restored palatial theater in Los Angeles, a contemporary of the better-known Mann's Chinese Theatre across the street. In a building next to the El Capitan, Disney had put up a small theme park based on the film, a three-story attraction featuring more than a hundred performers in several live shows, an obstacle course in which children attached their feet to green bases and hopped like the film's green army men, and a Pizza Planet eatery.

Catmull and two of his young children went to the premiere with Kerwin and another Pixar manager and his son. The six of them found that they had arrived early, so instead of going right to the theater, they decided to take a walk around the block. After they rounded a corner, they came upon something that made them freeze: a Burger King, seemingly stacked to the gills with all things *Toy Story.* There were *Toy Story* posters everywhere, *Toy Story* cups, *Toy Story* plates, little *Toy Story* figures to go with Burger King Kids Meals. For the adults in the group, the scene didn't just break the fourth wall, it left it in splinters. They already knew, abstractly, that Burger King had signed on, but nothing had prepared them for the actual sight of *their* Woody and *their* Buzz out loose in reality. It was then they realized that *Toy Story* was, in fact, no longer their film, and its characters were no longer their characters; Woody and Buzz now belonged to the world.

Pixar had a handful of passes to the premiere, but the guest list was otherwise Disney's. All the Disney film executives were on hand, as were Tom Hanks and some of the other voice actors, along with non–*Toy Story* celebrities such as Robin Williams. Buzz Aldrin was present at Pixar's invitation, since his namesake was one of the main characters. The audience appeared to be captivated by the film; adult-voiced sobs could be heard during the quiet moments after Buzz Lightyear fell and lay broken on the stairway landing.

By the time *Toy Story* released to the public several days later on more than twenty-four hundred theater screens, critics had already posted their judgments on the basis of advance screenings. All cynicism and contrarianism seemed to have been put aside for the occasion. Although the film had Disney's child-friendly imprimatur on it, the critics intuited that *Toy Story* was meant to connect with adults as much as children, and they embraced it on both levels. Janet Maslin of *The New York Times* hailed the film's "utterly brilliant anthropomorphism" and "exultant wit." *Time*'s Richard Corliss contrasted what he saw as the flatness of the characters in many of the day's live-action movies with the liveliness and imperfect hearts of Woody and Buzz, declaring *Toy Story* "the year's most inventive comedy." For David Ansen of *Newsweek,* it was a "marvel" that "harnesses its flashy technology to a very human wit" and to "rich characters." The widely syndicated television show *Siskel & Ebert* awarded it "two thumbs up." Owen Gleiberman of *Entertainment Weekly* reported, "I can hardly imagine having more fun at the movies than I did at *Toy Story.*" Kevin McManus of *The Washington Post* held, "For once, reality lives up to hype. With *Toy Story,* gigantic superlatives become appropriate, even necessary. . . . In fact, to find a movie worthy of comparison you have to reach all the way back to 1939, when the world went gaga over Oz."

There were a few discordant notes. Some critics, evidently confused by the prominent Disney branding, thought that the entire production was Disney's work or assumed that Pixar had done only the animation. The prevalence of male characters in the film led a *Chicago Tribune* critic (who admired the film in general) to complain that "*Toy Story* is so much of a boy movie that it almost seems like Disney's apology for *Pocahontas. . . .* Why does Andy's only girl toy have to be a prissy, fluttery-eyed throwback like Little Bo Peep?"

THANKSGIVING'S B.O. IS 'TOY' LAND
Disney film leads pack to holiday frame record

—*DAILY VARIETY* FRONT-PAGE HEADLINE, NOVEMBER 27, 1995

TOY BULLIES B.O. PLAYMATES
Film is first to amass $20 mil in a three-day Dec. weekend

—*THE HOLLYWOOD REPORTER* FRONT-PAGE HEADLINE, DECEMBER 5, 1995, REPORTING ON *TOY STORY*'S SECOND WEEK OF RELEASE

Pixar, formerly the ugly duckling of Silicon Valley, up for sale just a little more than a year earlier to anyone who would write Jobs a check for $50 million to cover his losses, now enjoyed a reversal of fortune. *Toy Story* was easily the U.S. box-office champion upon its release, with estimated receipts of $10 million going into the weekend (it premiered on a Wednesday) and another $28 million for the three-day weekend itself. It was the most successful Thanksgiving-weekend debut ever. In its first twelve days, *Toy Story* earned some $64.7 million altogether.

Although Thinkway had miraculously gotten its Woody and Buzz toys from design to production in just five and a half months, supplies quickly vanished from shelves. An executive at Hasbro, manufacturer of Mr. Potato Head (voiced in the film by Don Rickles), jokingly told Guggenheim that the company was experiencing a "potato famine."

"Quite honestly, we are totally amazed by how much good press our little movie has gotten," Pixar expressed on its simple Web site,

John Lasseter, following the 1995 release of *Toy Story,* with the film's stars. Disney's consumer products division gave little attention to Woody and Buzz while the film was in production; Disney nearly failed to license the characters to any toy company until a small Toronto-based firm stepped up and agreed to gamble on producing them.
© *Eric Robert/Corbis Sygma*

"and are just as surprised as anyone at the magnitude of the marketing and merchandising that has accompanied this film. You can bet, we all bought our Burger King Kid's Meals!!"

Jobs had ingeniously set Pixar's initial public offering day for November 29, a week after the opening of *Toy Story.* Technically, under SEC rules, Pixar was in a "quiet period" during which it could do nothing publicly to promote its offering. With the sudden omnipresence of *Toy Story,* Jobs wouldn't need to.

Catmull and Kerwin gathered that morning at the offices of Robertson, Stephens & Co., the lead underwriter for the offering. From the firm's sleek upper-floor quarters, one could look out over the city of San Francisco and the Bay. The bankers had Odwalla carrot juice, Jobs's favorite, on hand for a celebratory toast when the shares went on sale. What they didn't have was Jobs himself, who was late driving in. Those on hand began murmuring nervously, *Oh my God, he's going to miss the beginning of his IPO.*

Jobs came in just in time to see the shares disappear at the offering price of $22. Everyone watched, agog, as the shares escalated into the forties within an hour. As the day went on, trading would have to be delayed at times for a lack of sellers. The stock reached as high as $49.50 before closing at $39. After the bankers' fees were subtracted, the sale had raised $139.7 million for the company, nosing ahead of Netscape as the biggest IPO of the year.

Jobs had retained ownership of 80 percent of Pixar. At age forty, the former Atari technician vaulted out of the ranks of the merely wealthy. Following the IPO, his shares of Pixar were valued at more than $1.1 billion—and the *rounding error* on that figure was almost as much as the entire value of his Apple holdings when he left Apple a decade earlier.

Toy Story would go on to become the highest-grossing film of 1995, with $192 million in U.S. box-office receipts and $357 million globally. It would be the first animated film ever nominated for an Academy Award for best original screenplay. For Lasseter, there would be a special achievement Academy Award in 1996 "for his inspired leadership of the Pixar 'Toy Story' team, resulting in the first feature-length computer-animated film." (He arrived at the awards ceremony in a chauffeur-driven Oscar Mayer Wienermobile.)

Apart from revenues and accolades, Pixar and Disney gained much from *Toy Story.* Simply by making the first computer-animated film, Catmull, Lasseter, and others at Pixar had realized the fruition of a long-held dream. They had learned, moreover, that they could trust their own instincts and methods in the context of feature filmmaking—a self-assurance that comes with seeing a project through from an idea to a success. *Toy Story* gave validation to the view of Lasseter and his team that an animated feature could eschew fairy-tale plots and instead focus on adultlike characters with adultlike problems, while still providing entertainment to children; it was an approach that would recur time and again in later Pixar features. *Toy Story* also imprinted Pixar with a model of perfectionism and creative passion.

"There was such a palpable passion to make that film," Andrew Stanton remembered later. "I think every [Pixar] film since has just been an exercise in trying to re-create that passion."

At the same time, Pixar gleaned insights into Disney's ways in feature filmmaking.

"We learned considerably from Disney," Guggenheim recalled.

> On the organizational side, we learned how they worked budget and schedule together to keep a constant perspective on the state of the project. We also adopted some of their organizational hierarchy, but we kept it more "lean and mean" than they did, so we made do with far fewer people than they would on their projects. Creatively, we deepened our understanding of their discipline that the story is first and foremost. This had always been the mantra that John Lasseter preached at Pixar, so it wasn't a new idea, but we learned more about the depths to which they would take it. Feature films also have greater complexity of story, and this was an area where they knew the pitfalls better than we did.
>
> Ironically, though, several of their senior execs admitted to me by the end of the *Toy Story* production that Pixar had made a film

that contained more of the "heart" of traditional Disney animated films than they themselves were making at that time. They grudgingly admired Pixar and Lasseter for this.

Disney also learned from Pixar. Disney Feature Animation had already made adventurous use of computer animation here and there—to create the ballroom scene of *Beauty and the Beast,* the clock tower scene of *The Great Mouse Detective,* and the wildebeest stampede and other special effects in *The Lion King.* With *Toy Story,* however, Pixar opened Disney's eyes to the medium's breadth of possibilities.

It was, for the moment, too early to take seriously the possibility that the business landscape for Disney had just undergone a momentous change—that Disney might have unwittingly opened itself up to meaningful competition in animated feature films. It was a ridiculous idea. Upstarts like Don Bluth (*The Secret of NIMH, An American Tail, The Land Before Time*) had come and gone, for all intents and purposes. In the end, Disney still owned feature animation. It had always been so. It would always be so. Disney would dominate computer animation as it had dominated cel animation. Pixar would be the eager-to-please contractor. The stars and the planets seemed to be set in their courses—but the medium of computer animation, and the phenomenal success of *Toy Story,* was about to upend the old certainties.

8. "IT SEEMED LIKE ALL-OUT WAR"

A Bug's Life, Toy Story 2

During the summer of 1994, a little more than a year before the release of *Toy Story,* Pixar's story department began turning its thoughts to the next film. However appealing it might have been to devote the company's entire attentions to one film at a time, practicality dictated otherwise. Between Pixar itself and the Hi Tech Toons production company, there were roughly 160 employees on the payroll, scores of whom would become free as the stages of the production process for *Toy Story* wound up. At most other studios, those employees would simply be sent out the door until the next production. Catmull had rejected Hollywood-style run-of-show employment, believing that steady employment relationships would help the company hold on to its invaluable talent. Without run-of-show employment, however, Pixar needed to keep its artists and engineers productively engaged.

The crew members coming off *Toy Story* could be assigned to various smaller projects for a while. By and large, however, without a feature film to work on, the *Toy Story* crew would soon become so much costly overhead. The production pipeline that Pixar had created for *Toy Story* needed to be filled up again as rapidly as possible—which, in turn, meant having a Disney-approved story for the second film ready to go as soon as possible after *Toy Story* wrapped.

The story line of *A Bug's Life* originated in a lunchtime conversation between Andrew Stanton and Joe Ranft. Lasseter and his story team had already been drawn to the idea of insects as characters; insects, like toys, were within reach of computer animation at the time thanks to their relatively simple surfaces. Stanton and Ranft wondered whether they could find a starting point in Aesop's fable of the grasshopper and the ants.

In that story, a grasshopper squanders the spring and summer months on singing while the ants put food away for the winter; when winter comes, the hungry grasshopper begs the ants for food, but the ants turn him away. Disney had told its own version, one with a cheerier, more Disney-like ending, in the 1934 short *The Grasshopper and the Ants.* In the Disney film, the Queen of the Ants kindly allows the grasshopper to earn his meal by fiddling for it. As Stanton and Ranft discussed the fable, they hit on the notion that the grasshopper, the larger insect by far, could just *take* the food. They rattled off scenarios and story lines springing from that premise.

Lasseter liked the idea and offered some suggestions. The concept then simmered until early the next year, when the story team began work on the second film in earnest. That June, at an early test screening for *Toy Story* in San Rafael, they pitched the insect film to Michael Eisner. Eisner thought it was fine, and they submitted a treatment to Disney in early July under the title *Bugs:*

ACT I

End of Summer: A peaceful colony of black ants toils storing up food for the winter. An alarm is sounded, the ants quickly scurry for cover and hide the food they've been harvesting.

Suddenly, the sun is blotted out by a wave of grasshoppers that enter the colony like an unruly motorcycle gang. As the dust settles, Hopper, the leader of the gang, appears before the

> trembling ants. Once again, the grasshoppers have been playing all summer and neglected to save any food for the winter. "Lucky for us," says Hopper, "you busy little boy scouts have been hard at work. We'll do our usual deal. You harvest for the rest of the summer, we'll be back at the first winter frost to collect it, and we will let you live." The ants, a timid species, are unable to defend themselves, and Hopper and his horde fly away.
>
> Angry and frustrated, the ants hold a meeting with the wise old Queen Ant. She advises the nest to hire other bugs who are strong and clever enough to defend the ant colony from the evil grasshoppers. The Queen's daughter, Princess Atta, thinks this idea is too risky, and if it fails, will only further incur the wrath of Hopper. But she is overruled. Two ants are then chosen to leave the colony to scout for the greatest warriors in all bugdom. . . .

The leading man in this version of the story—or, rather, the leading arthropod—is a fire ant named Red, the slick-talking ringmaster of an insect circus. After the circus's owner, P. T. Flea, shuts it down for lack of attendance, the insects go to an insect bar to wash away their sorrows and figure out what to do. There, Red overhears the two ant scouts from the colony tell their story. He convinces the other circus insects to pretend to be tough, experienced fighters so they can land the job. They plan to live off the ant colony until winter, then sneak away before the grasshoppers come back.

Disney approved the treatment and gave notice on July 7, 1995, that it was exercising its option of a second film under the 1991 agreement. Lasseter assigned Stanton the job of co-director; the two men worked well together and had similar sensibilities. While making *Toy Story,* Lasseter had found that the workday of a sole director on a computer-animated feature was stretched to the point of fraying by everything that needed to be done, especially in the final twelve-month stretch. Masses of material throughout the pipeline needed to be evaluated in one review session after another. The stress was intense. A co-director would relieve some of the pressure, and Lasseter also felt the role would groom Stanton for a lead directing position of his own.

When *Toy Story* released that November, there was no script for *Bugs* yet, so after some rest and relaxation, the crew dispersed into other projects within the company. Some went to work animating a series of ten-second and thirty-second bits with the *Toy Story* characters for ABC Television, called *Toy Story Treats;* ABC aired them among Saturday-morning cartoons and commercials. Other, more technical people were drafted into the research-and-development department to improve some of Pixar's software tools based on the experience with *Toy Story.* One team worked on the lighting software to make it accessible to artists, so that it would no longer require a programmer to light a shot; another group worked on Menv to improve its handling of large shots with numerous models. Pixar's new interactive group under Pam Kerwin—which was creating CD-ROM games based on *Toy Story*—took on some people. Still others went to the story and art teams for *Bugs.*

Lasseter believed it would be useful to look at a view of the world from an insect's perspective. Two technicians obliged by creating a miniature video camera on wheels, which they dubbed the Bugcam. Fastened to the end of a stick, the Bugcam could roll through grass and other terrain and send back an insect's-eye outlook. Lasseter was intrigued by the way the grass, leaves, and flower petals formed a translucent canopy, as if the insects were living under a stained-glass ceiling. It was a look that he wanted for *Bugs.* Later, the team would also seek inspiration from *Microcosmos,* a French documentary on love and violence in the insect world.

After Stanton had completed a draft of the script, he came to doubt one of the story's main pillars—that the circus bugs who had come to the colony to cheat the ants would instead stay and fight. He felt the circus bugs were unlikable characters as liars and that it was unrealistic for them to undergo such a complete change of personality. Although production planning was already far along, Stanton concluded that the story needed a different approach.

Stanton's crisis of confidence was arguably needless. His story for

Bugs had the circus troupe growing to admire the ants, "impressed with their sense of community, family and hard work." Princess Atta, initially distrustful of Red, would begin to admire him for giving the ants the courage they had never had. Red was to delay the circus bugs' departure with one excuse after another because he had developed feelings for the princess. Finally, the morning of the first winter frost, the circus bugs would panic and start to leave, but too late; on their way out, they would run into Hopper and his gang. Red would be on the brink of revealing himself as a coward and explaining everything to Hopper—but would stop short when he glimpsed Atta watching him with pride. Red would then stand up to Hopper with blustery words, and the now confident ants would present a phalanx of resistance. When Hopper returned with his gang later, the circus bugs, sympathetic to the ants, would help to defeat the grasshoppers with acrobatics and their other circus stunts.

It was an engaging story, and one that dovetailed, in a way, with Buzz's evolution in *Toy Story.* Where Buzz had to reconcile himself to the disappointment of learning what he was—a toy, not a spaceman—Red was to find that if you put on a mask to look more noble than you are, you might just grow into the mask.

Stanton, in any event, rewrote the script to focus on one of the scout ants, the character who ultimately became Flik. Red disappeared from the story. The circus bugs, no longer out to cheat the colony, would be embroiled in a comic misunderstanding about what Flik was recruiting them for. Lasseter agreed with the new approach, and comedy writers Donald McEnery and Bob Shaw spent a couple of months at Pixar working with Stanton on further polishing.

The transition from treatment to storyboards took on an extra layer of complexity on account of the profusion of story lines. Where *Toy Story* focused heavily on Woody and Buzz, with the other toys serving mostly as sidekicks, this film required in-depth storytelling for several major groups of characters.

"That one was a really big struggle in the story department," remembered BZ Petroff, who was story supervisor on the film. "You had the circus bugs, you had Flik and the ant colony, and then you had the grasshoppers. And all of those characters needed to have great character arcs. At that time, we were trying to keep the film under eighty minutes—it's hard to tell that many stories in that short a period of time."

Character design also presented a new challenge: Where most people tend to find classic toys appealing, truly realistic insect characters would come across as creepy-crawly. Although the art department and animators studied insects closely, natural realism would give way to the larger needs of the film.

"We took out mandibles and hairy segmentation yet still tried to keep design qualities and aspects of texture that made you feel like you were looking at bugs," Stanton said. "We wanted people to like these characters and not be grossed out by them."

To make the ants more likable, the art department designed them to stand upright and replaced their normal six legs with two arms and two legs. The grasshoppers, in contrast, received a pair of extra appendages to make them *less* attractive.

Lasseter had come to envision the film as an epic in the tradition of David Lean's *Lawrence of Arabia.* The scale of the story required the software engineers to accommodate new demands. Among these was the need to handle shots with crowds of ants. The film would include more than four hundred such shots in the ant colony, some with as many as eight hundred ants. It was impractical for animators to control those ants individually, but neither could the ants remain static for even a moment without appearing lifeless. They could not move identically, either, without seeming mechanical and breaking the spell of the scene.

Bill Reeves, one of two supervising technical directors on the film, dealt with the quandary by leading the development of software for autonomous ants. The ants would take broad direction, as an actor would—delivering nervousness, joy, or whatever the shot required—but the software would make each ant react with slightly different behavior and timing. For the behaviors, Reeves relied on a group of animators to build a library of more than four thousand distinct motions that the software could draw upon. The ants thus appeared to respond as individuals. The crowd effects were similar in concept to Reeves's invention of a decade and a half earlier, particle systems, which had let animators use masses of self-guided particles to create effects like swirling dust and snow.

Other films had used autonomous characters in computer animation to depict crowds of animals, including the wildebeest stampede in *The Lion King* (1994), the swarming bats in *Batman Returns* (1992), and, seminally, the flying birds in the short *Stanley and Stella* (1987). Pixar's breakthrough was to do it with feeling—to give the autonomous characters the ability to project individual emotions.

The voice cast was heavy with television situation-comedy stars of the time: Flik was Dave Foley of *NewsRadio;* Princess Atta was Julia Louis-Dreyfus of *Seinfeld;* Hopper's obtuse brother was Richard Kind of *Spin City;* Slim, the circus's walking stick, was David Hyde Pierce of *Frasier;* Dim, the rhino beetle, was Brad Garrett of *Everybody Loves Raymond.* Lasseter cast Joe Ranft as Heimlich the caterpillar after Lasseter's wife, Nancy, heard him playing the character on a scratch voice track.

Only the casting of Hopper proved problematic. Lasseter's top choice, Robert De Niro, repeatedly turned the part down, as did a succession of other actors. The cause of the string of refusals was never clear; some in the company wondered whether sentimentality made De Niro and the others reluctant to play a scary-looking character in a film that children would see. In the end, Kevin Spacey took the part and gave Hopper an icily confident villainy.

From a business standpoint, the film would inaugurate a new relationship between Pixar and Disney. A few months after the release of *Toy Story,* Jobs had begun pressing Eisner to supplant the 1991 three-film deal with a radically different contract. Where Pixar was cash-strapped at the time of the 1991 deal, and willingly agreed to a mere 10 to 15 percent share of the films' profits in return for Disney's full financing of the films, Pixar now had the cash (thanks to the initial public offering) to co-finance its films and insist on a higher percentage. Of no less importance, Jobs was unhappy that many consumers believed *Toy Story* had been a Disney-made film; he wanted contractual guarantees that Pixar's name would appear more prominently.

Jobs had something of value to offer Eisner in exchange: more films. It was enough. On February 24, 1997, Pixar's chief financial officer, Lawrence Levy, and his Walt Disney Pictures & Television counterpart, Robert Moore, signed a forty-two-page contract for a new five-film deal; *Bugs,* as it was still called, would be the first. Production costs would be shared fifty–fifty, and Pixar would receive 50 percent of Disney's receipts (after distribution costs) from box office, home video, and tie-in products such as clothes and toys. Pixar would get equal billing with Disney in its films, in advertising, and on tie-ins. Pixar's stock, which had slumped to fourteen dollars at this point, went up 50 percent the day that Jobs and Eisner announced the contract.

In his annual letter to shareholders that summer, Jobs explained that the more prominent branding he had obtained in the contract was part of an ambitious vision for the company—no less than to establish Pixar in the minds of parents as another Disney:

> We believe there are only two significant brands in the film industry—"Disney" and "Steven Spielberg." We would like to establish "Pixar" as the third. Successful brands are a reflection of consumer trust, which is earned over time by consumers' positive experiences with the brand's products. For example, parents trust Disney-branded animated films to provide satisfying and appropriate family entertainment, based on Disney's undisputed track record of making wonderful animated films. This trust benefits both parents and Disney: it makes the selection of family entertainment that much easier for parents, and it allows Disney to more easily and assuredly draw audiences to see their new films. Over time we want Pixar to grow into a brand that embodies the same level of trust as the Disney brand. But in order for Pixar to earn this trust, consumers must first know that Pixar is creating the films.

At the same time Jobs was repositioning Pixar as an equal partner to Disney, he was enjoying another, equally improbable turn of events with a triumphant return to Apple Computer. His route back was roundabout. Apple executives had concluded that the company needed to look outside its own walls for the next generation of the Macintosh operating system. Among the options Apple considered were licensing Microsoft Windows or buying Jobs's NeXT for its advanced software, known as OPENSTEP. In the end, Apple bought Jobs's company for $427 million, a decision announced in December 1996. Apple's then-CEO, Gil Amelio, brought in Jobs as a part-time adviser, figuring that Jobs would serve as "the front guy," as Amelio remembered, "this glamour boy out there keeping the faithful in line."

Amelio had misjudged his new counselor. Jobs quickly won the confidence of an influential member of Apple's board, Edgar Woolard, Jr., a former chairman and CEO of DuPont. Amelio was out of his job the following July and Jobs became "interim CEO" on September 16, 1997. He drew a salary of one dollar a year and refused to accept any stock options.

The change was a relief for senior people at Pixar, who were not displeased to see Apple taking up more of Jobs's attentions.

"Steve's so smart and has so much to give and so much energy that he needed to put it somewhere," Kerwin recalled.

> And there was a little bit of tension in that he had more to give Pixar than Pixar actually wanted. . . . So when he went back to Apple, it was great because that could absorb his passion, and Steve could still do the stuff that Pixar really needed Steve to do, which was managing Disney, getting Pixar better distribution terms.

Jobs's deal making had secured a place for Pixar side by side with Disney in the firmament of animation. Both Pixar and Disney now had an acrimonious challenger, however, in the person of Jeffrey

Jeffrey Katzenberg with the DreamWorks Animation character Shrek, at the Australian premiere of *Shrek 2,* June 9, 2004, Sydney.
Patrick Riviere/Getty Images

Katzenberg. After leading the revival of Disney animation, Katzenberg had lost his job and had left in a state of anger toward Eisner, his erstwhile mentor and boss. Although Katzenberg had given Pixar its start in feature films a half-dozen years earlier, he was about to make *Bugs* the target of an attack by proxy on Disney. The schism had been set in motion several years earlier by a heli-skiing accident in the remote Ruby Mountains of eastern Nevada.

During Easter weekend 1994, the president and chief operating officer of the Walt Disney Co., Frank Wells, hosted a skiing trip for a few old friends—his mountain-climbing comrade Dick Bass, the actor Clint Eastwood, and a husband-and-wife team of adventure-documentary filmmakers, Mike Hoover and Beverly Ann Johnson, as well as one of Wells's grown sons. Both Wells and Eisner had been leading candidates for the CEO position at Disney in 1984, until Wells gracefully settled the issue by agreeing to take the number two job. Since joining Disney, Wells, a Rhodes scholar, had become known for his even-keeled discretion, his unassuming manner, his diplomacy, and his consideration, traits that were not overabundant in Eisner himself.

On the afternoon of Sunday, April 3, the last day of the trip, news of a fast-moving snowstorm led the skiers to cut their afternoon short. A pair of Bell 206 helicopters from Ruby Mountain Heli-Ski flew in to pick them up at around two o'clock and ferry them back to the lodge where they had been staying. The plan was for the skiers to grab their bags at the lodge, then reboard the helicopters to fly to an airstrip where a waiting Walt Disney Co. jet would take them to Los Angeles.

Wells, Hoover, and Johnson volunteered to wait for the second helicopter so that the other guests wouldn't have to worry about becoming stranded. Wells's son, Kevin, offered to wait with him. Wells reminded Kevin that they had promised Wells's wife, Luanne, that they wouldn't fly in the same helicopter during the weekend.

But it's just a short hop, Kevin pointed out.

"No, that's what we agreed to," Wells said with finality. "I'll stay and you go."

A few minutes after the first helicopter left with Bass, Eastwood, and Wells's son, the second one arrived. Wells, Hoover, and Johnson loaded their skis in a hurry and lifted off. With them inside were their skiing guide, Paul Scannell, and the pilot, David Walton. Walton quickly saw fog closing in from all directions. They had run into the leading edge of the storm. He brought the helicopter down on a knoll and waited for the weather to pass.

After about an hour, Walton warmed up the engine for a while and cleared snow from the top of the helicopter. A little past four-thirty, the weather finally cleared and Walton took the party up again.

Within minutes, several lights went red along the top of the helicopter's instrument panel and the passengers heard the beeping of an alarm. The engine had cut off. The pilot calmly radioed that they had engine failure.

As the helicopter started to drop, Hoover urged his wife and Wells to brace themselves. His wife did so. Wells sat still and began asking questions, as if the situation were an intellectual problem that could be solved with some fact-finding and cogitating.

"*Brace yourself,*" Hoover repeated.

An instant before the helicopter hit ground, the pilot uttered a *whoa*—as if to say, *This is going to be bad.*

David Walton and Frank Wells died in the crash, and Beverly Ann Johnson died minutes afterward. Paul Scannell, the guide, died from his injuries a week and a half later. Mike Hoover, the sole survivor, was a mass of broken bones.

The National Transportation Safety Board later attributed the crash to snow in the engine, which may have accumulated while the helicopter was parked in the storm. The NTSB noted that the helicopter's Allison engine was known to stall after taking in small amounts of snow.

In Burbank on Monday, Katzenberg was shocked to learn from a press release that Eisner would assume Wells's former titles. Katzenberg was saddened by Wells's death, but he had understood Eisner to have promised *him* Wells's job if Wells left. Katzenberg, now forty-three, had spent the past nineteen years working for Eisner at Disney and, before that, at Paramount—nearly half his life. Katzenberg had staked his career on the belief that Eisner valued his loyalty, and he took for granted that Eisner would treat him as his protégé. All his prodigious work and accomplishments for Eisner suddenly seemed to turn to ashes.

Over lunch with Eisner the next day, he demanded to know where he stood. As Eisner sidestepped the issue, the conversation deteriorated and Katzenberg hinted he was ready to leave the company if he didn't get the promotion.

"I am angry because one day after Frank's death he [Katzenberg] had the bad taste and audacity to demand Frank's responsibilities and title or else he, Jeffrey, would leave the company," Eisner wrote to his lawyer a couple of weeks later, for the sake of getting his thoughts onto paper. "Frankly I do not believe there are so many jobs open right now although I will not kid myself. Jeffrey will do very well. He will end up a high level executive, but eventually he will fail."

Eisner made up his mind to fire Katzenberg, but would do so on his own schedule, when he had his ducks in a row. He assured Katzenberg later in the day that he had an excellent future at Disney and let him go four months afterward.

On October 12, Katzenberg—together with partners Steven Spielberg and David Geffen—announced the formation of a new studio, soon to be named DreamWorks SKG. (The initials were those of the founders.) At the press conference that morning at the Peninsula Hotel in Beverly Hills, the men were sketchy about their intentions, but vowed to compete hard with Disney in animation. Katzenberg disclaimed any lingering animosity toward Eisner.

"Nineteen years with Michael Eisner was the most extraordinary experience of my life," Katzenberg said. "He was my mentor, teacher, friend. I truly would not be at this table today were it not for him. I don't think either of us are thrilled to death about the hurt, but that hurt is behind us."

Roy E. Disney, who took over responsibility for animation at Disney after Katzenberg's departure, was publicly unconcerned by the turn of events. "This is not the first time others have said they'd get into the animation business," he told *The New York Times* several days later. "It takes more money and time than most have been willing to spend."

Lasseter and Katzenberg kept in touch. Lasseter remembered well that Katzenberg had led the charge at Disney for a Pixar film deal at a time when executives at other studios, and even at Disney itself, were uninterested in computer animation. Without him, Lasseter knew, there probably would have been no *Toy Story.* Beyond that, Lasseter respected his judgment and felt comfortable using him as a sounding board for creative ideas.

While Lasseter was in Los Angeles in October 1995 to oversee postproduction work on *Toy Story,* he dropped by Katzenberg's office and excitedly told him about *Bugs.* As Lasseter recalled the conversation, Katzenberg's main reaction was to query him on when *Bugs* was coming out.

At the same time, Katzenberg was shopping for a computer graphics studio. (He meant what he said about competing with Disney.) What was available were houses that specialized in special effects or commercials. Several had achieved particular prominence: White Plains, New York–based Blue Sky Studios, home of the director Chris Wedge, who had made two well-regarded short films; Rhythm & Hues Studios of Los Angeles, which had animated the expressive animal faces in *Babe;* and Pacific Data Images of Sunnyvale, California, which had worked on Michael Jackson's "Black or White" music video, television commercials for Pillsbury and other national clients, a 3-D version of Homer Simpson for an episode of *The Simpsons,* and special effects for films such as *Batman Forever* and Disney's *Angels in the Outfield.*

Katzenberg settled on Pacific Data Images, or PDI, and announced DreamWorks' purchase of 40 percent of the studio in early March of 1996. As it happened, Pixar and PDI had practically been sibling companies. PDI started in 1980, around the same time as the Computer Division of Lucasfilm, and people from the two Bay Area firms had long been friends and partied with one another. The Pixar–PDI beach party at the 1988 SIGGRAPH was merely the most visible expression of the companies' easiness with each other. After Pixar entered the television commercial business in 1989 and became a competitor of PDI's, the relationship remained friendly; the two companies even sent clients to each other when they faced an overflow in their workloads. As pioneers in a still-nascent field, the leaders of the two companies believed that the other company's successes would strengthen the industry as a whole and ultimately benefit them both. PDI's founders, Carl Rosendahl, Richard Chuang, and Glenn Entis, remained tolerant as Pixar hired people away from PDI to make *Toy Story.*

After the acquisition, Lasseter and others at Pixar were dismayed to learn from the trade papers that PDI's first project for Dream-

Works would be another ant film, to be called *Antz.* Katzenberg had heard a pitch for the film from one of his executives, Nina Jacobson, and decided it would be an excellent idea. By this time, Pixar's *Bugs* project was well known within the animation community.

Lasseter recalled phoning Katzenberg and remonstrating with him, "Jeffrey, how could you?"

"He started talking about all this paranoid stuff about conspiracies—that Disney was out to get him," Lasseter recounted. "He said he had to do something. That's when I realized it wasn't about me. We were just cannon fodder in his fight with Disney."

In truth, Katzenberg *was* the victim of a conspiracy: His former boss had decided not to pay him a dime of the bonus (amounting to more than $100 million) that he was owed under his Disney contract, and had convinced Disney's board not to give him anything. In the rough-and-tumble of film industry competition, Eisner had also scheduled *Bugs* to open the same week as *The Prince of Egypt,* which was then intended to be DreamWorks' first animated release.

Soon afterward, at a regular company-wide Friday meeting at Pixar, Lasseter stood in front of the room with Catmull and Jobs and grimly relayed the news. Both Lasseter and Jobs then told the assembled Pixar employees not to be distracted by the developments. We're going to make the best film ever, they said, and the best film is the one that will win. Privately, Lasseter told other Pixar executives that he and Stanton felt terribly let down because they had regarded Katzenberg as a supporter and a kindred creative spirit.

Katzenberg was not finished yet. Word eventually made its way to Point Richmond that he would not only be making a similarly themed film, he would be releasing it almost two months ahead of *Bugs,* moving the opening of *Antz* from March 1999 to October 1998—an obvious attempt to undermine Pixar's release. The rumor on the street, never confirmed, was that Katzenberg had given PDI rich financial incentives to induce them to do whatever it would take to have *Antz* ready first, despite Pixar's head start. *Antz* would now be DreamWorks' first animated feature.

"What we were hearing was that . . . they had to get this movie done in a short time and that it didn't matter whether it was all that good or not," Kerwin said. "It seemed like all-out war at the time."

Jobs and Lasseter later made the charge—denied by Katzenberg—that the DreamWorks head phoned each of them and offered a proposition: If they would persuade Disney to change the release date of *Bugs,* he would shut down work on *Antz.* Lasseter said he hung up on him.

As production continued on *Bugs,* shortly to be renamed *A Bug's Life,* the crew took little notice of the war of the ant films. Although the contention left Katzenberg and Lasseter estranged, Pixar and PDI employees kept up the old friendships that had arisen from spending a long time together in computer animation.

Lasseter and Stanton had two supervising animators to assist with directing and reviewing the animation, Rich Quade and Glenn McQueen. (McQueen, coincidentally, was one of the employees Pixar had recruited from PDI.) The first sequence to be animated and rendered was the circus sequence that culminated with P. T. Flea's "Flaming Wall of Death." Like the army-men sequence of *Toy Story,* the circus sequence was first in the pipeline because Lasseter judged that it was the least likely to change, whatever else might happen to the story.

The animators found it a joy to work with Lasseter again. In another respect, however, *A Bug's Life* was a step backward from the animators' point of view. Production was harder and more tedious for them than on *Toy Story* because the models moved sluggishly on their computers. The root of the problem was that the models of the insect characters were far more complex than the toy characters, and there were more of them. The modeling team on *Toy Story* had the luxury of time to optimize Woody's and Buzz's models, which meant the animators' workstations could quickly generate the simple preview frames that animators relied on. The crush of work on *A Bug's Life* left no time for tweaking the models under the hood, so every little step took the animators longer to do—every body movement, every facial expression.

The surfaces of the characters in *A Bug's Life* were more lifelike than earlier Pixar creations, thanks to a technique that Catmull had developed with Jim Clark during their graduate student days at the University of Utah. The technique, known as subdivision surfaces or subdivision meshes, was a way of describing and rendering three-dimensional surfaces; it made it easier to create surfaces with a soft, smooth, supple appearance. Older methods—like the bicubic patches Catmull had explored in his doctoral thesis—allowed for realistic images of many inorganic materials, such as plastic; they tended, however, to make skin (whether human or insect) look like . . . plastic. Subdivision surfaces offered more realistic skin and better control over surfaces with rumples and folds, like clothes.

A small team at Pixar worked out the practical problems that had been keeping subdivision surfaces out of animation. Before the newly reborn method took its place in *A Bug's Life,* Catmull had asked for a short film to test and showcase it. The result was the 1997 short *Geri's Game,* four and a half minutes in length, in which a histrionic old man plays chess against himself.

As the release dates for *Antz* and *A Bug's Life* approached, Disney executives concluded that Pixar should deal with the DreamWorks battle by keeping silent on it—there was nothing to be gained, as they saw things, by talking about another studio's film. Jobs, however, had difficulty simmering in peace; over Disney's objections, he disparaged Katzenberg's tactics in a series of press interviews. "The bad guys rarely win," he told the *Los Angeles Times.* ("Steve Jobs should take a pill," DreamWorks' head of marketing, Terry Press, volleyed back.) Lasseter publicly dismissed *Antz* during this time as "a schlock version" of Pixar's film.

"It's sad, because they clearly stole the idea from us," Lasseter told an interviewer. "But we haven't worried about that too much. We've put it behind us."

Antz premiered at the Toronto Film Festival on September 19, 1998, and had its general release on October 2. Lasseter told others that if DreamWorks and PDI had made the film about anything other than insects, he would have closed Pixar for the day so the entire company could go see it. As it was, he claimed not to have seen the film himself.

A Bug's Life followed on November 25; it was initially set for November 20, but Disney pushed it back to avoid a collision with another competing film, Paramount's *The Rugrats Movie.*

Both *A Bug's Life* and *Antz* centered on a young male, a drone with oddball tendencies who was struggling to fit into a conformist ant society, and who would ultimately win a princess's hand by saving that society. Nonetheless, the differences were unmistakable. For humor, *A Bug's Life* relied chiefly on visual gags, while *Antz* was more verbal. The Pixar film was visually richer overall by a wide margin. The script of *Antz* was heavy with adult references, while that of *A Bug's Life* was more accessible to children. Also, thanks to Katzenberg's deep Hollywood network, *Antz* featured a long list of celebrity actors: The Flik-esque main character, Z, was voiced by Woody Allen; the ant queen was Anne Bancroft (her counterpart in *A Bug's Life* was Phyllis Diller); the ant princess was Sharon Stone. Others on the *Antz* marquee included Danny Glover, Gene Hackman, Jennifer Lopez, and Sylvester Stallone.

Despite *Antz*'s star power, and despite its having been at the front of the line, the insect war proved to be a rout. *A Bug's Life* grossed $163 million in its domestic theatrical release, 80 percent more than *Antz*'s $90 million, and $358 million worldwide, well over twice *Antz*'s $152 million. *A Bug's Life* did not do quite as well as *Toy Story* had domestically, but it matched its predecessor worldwide. Lasseter had been right: If Pixar concentrated on making an entertaining film, it would have nothing to worry about.

All the same, there was no need to make the game easier for the competition than it had to be. After *A Bug's Life,* Lasseter and his team would keep Pixar's films quiet until work was far along.

Still another film was in Pixar's pipeline during the making of *A Bug's Life.* Talk of a sequel to *Toy Story* began around a month after *Toy Story* opened, when Catmull, Lasseter, and Guggenheim visited

Joe Roth, Katzenberg's successor as chairman of Walt Disney Studios. Roth was pleased and embraced the idea.

Disney had recently begun making direct-to-video sequels to its successful feature films, and Roth wanted to handle the *Toy Story* sequel this way, as well. A direct-to-video sequel could be made for less money, with lesser talent. It could be priced cheaply enough to be an impulse purchase. Disney's first such production, an *Aladdin* spin-off in 1994 called *The Return of Jafar,* had been a bonanza, returning an estimated hundred million dollars in profits. With those results, all self-restraint was off; Disney would soon grace drugstore shelves with *Beauty and the Beast: The Enchanted Christmas; Pocahontas II: Journey to a New World; The Lion King II: Simba's Pride;* and still another *Aladdin* film.

Everything else about the *Toy Story* sequel was uncertain at first: whether Tom Hanks and Tim Allen would be available and affordable, what the story's premise would be, even whether the film would be computer-animated at Pixar or cel-animated at Disney.

As with *A Bug's Life,* Lasseter regarded the project as a chance to groom new directing talent. In early 1996, once Roth decided that Pixar would handle production of the sequel, Lasseter assigned directing duties. Stanton was immersed in *A Bug's Life;* Pete Docter, whom Lasseter regarded as the next in line, was already beginning development work on his own feature about monsters. For *Toy Story 2,* Lasseter turned to Ash Brannon, a young directing animator on *Toy Story* whose work he admired. Brannon, a CalArts graduate, had joined Pixar to work on *Toy Story* in 1993.

The story originated with Lasseter pondering what a toy would find upsetting. In the world of *Toy Story,* a toy's greatest desire is to be played with by a child. What, Lasseter wondered, would be the opposite of that—worse, even, than being displaced by another toy?

An obsessive toy-collector character had appeared in a draft of *Toy Story* and was later expunged. Lasseter felt that it was now an idea whose time had come. Thinking of his own tendency to shoo his sons away from the toys on his office shelves, especially a Woody doll that he prized for its Tom Hanks signature, Lasseter began talking about the notion of a toy collector who hermetically seals toys in a case where they will never be played with again. For a toy, it would be a miserable fate. Brannon then suggested the idea of a yard sale where the collector recognizes Woody as a rare artifact and distracts Andy's mom to grab him. Out of those ideas, *Toy Story 2* was born.

The concept of Woody as part of a collectible set came from the draft story of *A Tin Toy Christmas,* in which Tinny was part of a set in a toy store and became separated. The other characters in Woody's set emerged from viewings of 1950s cowboy shows for children, such as *Howdy Doody* and *Hopalong Cassidy.* "We started looking at these canonical characters that you find in westerns," said Guggenheim, who was producer of *Toy Story 2* during the first year of development work. "You would find a gruff old prospector. You would find other characters, like an Annie Oakley–Calamity Jane sort of character, a tough frontier girl."

The development of the cowgirl character, Jessie, was also kindled by Lasseter's wife; Nancy had pressed him to include a character in *Toy Story 2* for girls, one with more substance than Bo Peep. Jessie had started in a different form, as Señorita Cactus, a Mexican sidekick to the Prospector; she was to sway Woody with her feminine wiles. When the character of Jessie replaced her, the personality of the female lead became tougher and more direct.

As the story approached the production stage in early 1997, there remained the question of where Pixar would find the people to make it, given the demands of *A Bug's Life* on the company's employees. Part of the answer would come from a production organization within Pixar devoted to computer games. The Interactive Products Group, with a staff of around ninety-five (out of Pixar's total staff of three hundred), had its own animators, its own art department, and its own engineers. Under intense time pressure, they had put out two successful CD-ROM titles: *The Toy Story Animated StoryBook,* released in April 1996, and *The Toy Story Activity Center,* released in October of the same year to coincide with the videotape release of *Toy Story.* The games featured much of the voice cast of the film, except that the voice actor Pat Fraley took Tim Allen's place as Buzz, while Woody was played by Tom Hanks's younger brother, Jim. The company touted *StoryBook* as the first CD-ROM to deliver full-screen, motion-picture-quality animation on home computers. Between the two products, the interactive group had created as much original animation as there was in *Toy Story* itself.

Jobs had convinced himself that the games would sell ten million copies, like best-selling direct-to-video films. Kerwin, as head of the group, insisted that the market wasn't there on such a scale. We can make a good, profitable business out of them, she said. (The products had sold almost a million copies combined.) But they won't be a home run like *Toy Story.*

If that's the case, Jobs said finally, then why don't we just turn all these people over to making another movie? Thus, in March 1997, while Kerwin took the assignment of building a short-films group, Jobs shut down the computer games operation and the games staff became the initial core of the *Toy Story 2* production team.

Press Release

TIM ALLEN AND TOM HANKS RETURN AS "BUZZ LIGHTYEAR" AND "WOODY"

MARCH 12, 1997

The Walt Disney Studios and Pixar Animation Studios announce today that a sequel to the groundbreaking Academy Award–nominated feature film TOY STORY is underway and being created exclusively for home video. The all-new, fully computer-animated sequel will feature the voices of Tim Allen and Tom Hanks, who reprise their enormously popular roles as "Buzz Lightyear," the space ranger, and "Woody," the pull-string Cowboy, respectively. Production on TOY STORY II re-teams Disney's Feature Animation team and Pixar's Northern California studios. . . .

" 'Toy Story II' is the latest production to be announced in our growing made-for-video film category," Ann Daly, President, Buena Vista Home Video, said. "With 'Aladdin and the King of Thieves' and the debut of 'Honey We Shrunk Ourselves' next week, we are now bringing both animated and live-action films into this pipeline with great success. . . ."

Disney soon became unhappy with the pace of the work on the film and demanded in June that Guggenheim be replaced as producer. Pixar complied.

He looked back on his seventeen years with Pixar and Lucasfilm and concluded that he had most enjoyed working with groups that were venturing into new directions, like the EditDroid digital editing project and the original *Toy Story* effort; with Pixar's shedding of everything but feature films, he believed the company's strategy left few entrepreneurial opportunities. Guggenheim, now financially secure thanks to the stock offering, left the company.*

Karen Jackson and Helene Plotkin, who had been associate producers on the sequel, moved up to the role of co-producers. Jackson recalled using the enticement of greater responsibility—the chance to be a big fish in a smaller pond—to compete with *A Bug's Life* for the production people they wanted.

"You could go to *A Bug's Life* and be one of two hundred, or you could come to *Toy Story 2* and be one of fifty or sixty," she said. "To fill the spots on *Toy Story 2,* we did a lot of recruitment outside. But there were certain key positions on *Toy Story 2* where we wanted to get experienced staff on board, and the way to get them on board was to say, 'We'll let you run this department,' or, 'We'll let you be the directing animator.' "

In November, Disney executives Roth and Peter Schneider viewed story reels for the film, with some finished animation, in a screening room at Pixar. They were impressed with the quality of the work and became interested in releasing *Toy Story 2* in theaters.

In addition to the unexpected artistic caliber, there were other reasons that made the case for a theatrical release more compelling. As it turned out, the economics of direct-to-video for a Pixar film weren't working as well as hoped. The logic of direct-to-video hinged on low production costs, but low-budget and high-budget projects could not readily coexist under Pixar's umbrella. The cre-

* He subsequently went to the game company Electronic Arts, where he headed content development for the online game *Majestic* (2001), and later co-founded an independent animation studio, Alligator Planet.

ative appetites of Pixar's leadership made it anathema to produce a film at less than the highest level visually—one in which corner-cutting could be seen on screen. In computer animation no less than in live action, production values cost money.

More prosaically, Pixar wanted the efficiency of moving crew members from one production to the next, whatever the next one might be, so Catmull and Lasseter deemed it unacceptable to create a second, lower-wage staff for low-budget projects. Since labor costs added up to 75 percent or more of the production costs, it was unrealistic to try to make a significantly lower-cost production as long as all the films were to come from the same pool of employees earning the same salaries.

Lastly, animation salaries had gone up across the board. For *Toy Story,* Pixar had been able to hire people relatively cheaply on account of the excitement of working on a milestone in animation, the first fully computer-animated feature. Pixar remained attractive to potential hires by virtue of *Toy Story*'s quality and Lasseter's reputation, but Pixar also had new competition for talent; not only was DreamWorks producing traditional and computer-animated features, other studios were opening their own animation units following the success of *The Lion King.*

After some negotiating, Jobs and Roth agreed that the split of costs and profits for *Toy Story 2* would follow the model of the new five-film deal of 1997—but *Toy Story 2* would not count as one of the five films. Disney had bargained in the contract for five original features, not sequels, thus assuring five sets of new characters for its theme parks and merchandise. Jobs gathered the crew and announced the change in plans for the film on February 5, 1998.

Lasseter would remain fully preoccupied with *A Bug's Life* until it wrapped in the fall. Once he became available, he took over directing duties and added Lee Unkrich as co-director. Unkrich, who had just come off *A Bug's Life* as supervising editor, would concentrate on layout and cinematography. Brannon would also be credited as a co-director.

Up to then, the *Toy Story 2* team had been on its own—not just figuratively, but literally, having been placed in a new building that was well separated from the rest of the company by railroad tracks. "We were the small film and we were off playing in our sandbox," Jackson said.

That was about to change.

To make the project ready for theaters, Lasseter would need to add twelve minutes or so of material and strengthen what was already there. The extra material would be a challenge, since it could not be mere padding; it would have to feel as if it had always been there, an organic part of the film.

Unkrich, concerned about the dwindling amount of time left, asked Jobs whether the release date could be pushed back. There was too much to be done.

"I was sitting in Steve Jobs's office and I said, 'I'm utterly convinced that we can make a great movie here. I just don't think we can do it in the amount of time that we have,' " he remembered later.

"Steve basically said, 'Well, we have no choice. There are too many things lined up' "—presumably in reference to the film's licensees and marketing partners, which were getting toys and promotions ready.

Jobs buoyed Unkrich's spirits. "When I look back on my career," Jobs told him, "it's the things that were made under these circumstances, under these conditions that were not the best, that I'm the most proud of."

With the scheduled delivery date less than a year away, there was no time for months of noodling over the story. Lasseter called Stanton, Docter, Ranft, and some Disney story people to his house, a half-dozen blocks from Sonoma's nineteenth-century town square, for the weekend. There, he hosted a "story summit," as he called it—a crash exercise that would yield a finished story in just two days. Back at the office that Monday, Lasseter assembled the company in a screening room and pitched the revised version of *Toy Story 2* from beginning to end.

"Everybody was totally entertained," animator Mark Oftedal remembered. "It was a dramatic turn of events for the movie. It became something that was great and that everybody wanted to get in there and start animating."

The summiteers found parts of the story in ideas that Pixar had discarded from *Toy Story.* The opening sequence of *Toy Story,* at one point, was to be a Buzz Lightyear cartoon showing on television. Lasseter had dropped it from *Toy Story* in favor of a sequence showing Andy's relationship with Woody, but the concept of a Buzz Lightyear cartoon evolved into the Buzz Lightyear video game that would open *Toy Story 2.* In a draft of *Toy Story,* Woody was to suffer a nightmare after Buzz displaced him as Andy's favorite toy—a nightmare in which Andy was to throw Woody into a trash can and Woody was to become covered with hundreds of crawling roaches. That nightmare, in a milder form, would appear instead in *Toy Story 2* as a device for showing Woody's fear of rejection after his seam rips. The idea of a squeak-toy penguin with a broken squeaker also resurfaced from an early version of *Toy Story.*

"John has got a real eye for story," said Floyd Norman, a veteran Disney artist who had worked there during Walt's day, starting with *Sleeping Beauty,* and whom Pixar hired for story work on both versions of the *Toy Story* sequel.

> He [Lasseter] came in with a fresh eye and gave the film a nudge forward and raised the bar a little higher. I think our good film became a great film under John's direction.
>
> He'd see things like Woody becoming too much of a jerk in a scene. I remember we had a sequence in the film where Woody dreams of being lauded as this valued collectible, where he's a big shot. He's arriving by limousine and people are taking photographs. He's imagining this fame and hero worship. John looked at that and said, "Yeah, it's a funny idea, but it makes our character not as likable because now he's even more vain." And he cut that sequence.

Other changes included the luggage-belt chase scene at the airport, which became a bigger set piece than in the direct-to-video version; the concept of having the Prospector still in the box ("mint condition, never been opened"); and the addition of Jessie's song to tell her haunting story. Lasseter also looked closely at every shot that had already been animated and called for tweaks throughout—a different expression on a character in this shot, a different camera angle or lighting setup in that shot, a handful of extra frames at the end of another.

The film reused digital elements from *Toy Story,* the making of which had left behind a kind of digital backlot. The company's prevailing culture of perfectionism meant that it reused less of *Toy Story* than might be expected, however. The character models looked the same, but received major upgrades internally.

"The characters now interact better with themselves, each other and other objects in the film," modeling supervisor Eben Ostby said at the time.

> Their heads intersect with their chests better, which allows us to do more close-ups of the characters. On the first film, you couldn't really pull the mouth open and make a round shape. It always had a little tuck in the corners. Buzz and Woody now have more facial controls.

The shaders—the programs determining the appearances of surfaces—also went through revisions to bring about subtle improvements. For instance, Woody's plaid shirt and blue jeans looked the same in *Toy Story* and *Toy Story 2* from a distance, but in close-ups, one could see in *Toy Story 2* that they looked more cloth-like in their textures.

The *Toy Story 2* team did, however, freely borrow models from other productions. The terrain of the alien planet in the rapid-fire Buzz Lightyear video game was a modified version of Ant Island from *A Bug's Life;* during Jessie's song, the tree from which Jessie and her owner swing on a tire swing was also from Ant Island; the character of the Cleaner was Geri from *Geri's Game.* As Hamm the piggy

©DISNEY/PIXAR. COURTESY PHOTOFEST

FINDING NEMO (2003)

Pixar's fifth feature was visually its most lush up to that point—a contrast with its unusually dark ubject: a father clownfish, Marlin, whose wife and children have been killed, has become separated from his one surviving child. The film evoked the worst worries of parents. The company's brand was strong enough by then to enable it to take such risks.

The film drew upon highly detailed research on ocean life, though director Andrew Stanton opted to take some liberties for the sake of the story. Among them: in reality, male clownfish change sexes when the dominant female dies. Marlin, nonetheless, remained a male.

©DISNEY/PIXAR. COURTESY PHOTOFEST

THE INCREDIBLES (2004)

Writer-director Brad Bird crafted each of the four lead characters with enough depth and real-life resonances that a spectrum of moviegoers—adults and children—could feel that the film had been made just for them. Bird used the superhero genre to explore tensions of family life and male midlife crisis—without sacrificing action-movie thrills. Bird had created the story before coming to Pixar in 2000, conceiving it as a 2-D animated film.

As it happened, the Walt Disney Co. shut down much of its operations in 2-D feature animation the same year *The Incredibles* was released, reflecting then-CEO Michael Eisner's belief that audiences were no longer interested in traditional animation.

CARS (2006)

Although arguably Pixar's weakest film from an adult perspective, and its worst-reviewed, *Cars* offered a child-friendly script and technical virtuosity. It was the second-highest-grossing film of the year. As with the *Toy Story* films, the subject matter of *Cars*—NASCAR racing and a down-at-the-heels small town—emerged from director John Lasseter's personal interests.

The protagonist, "Lightning" McQueen, was named for Pixar animator Glenn McQueen, who died of cancer before the film's production began.

RATATOUILLE (2007)

Pixar's eighth feature, the story of a rat who wants to become a chef, was originally to be the studio's first outside its distribution setup with Disney. Jan Pinkava, who conceived the story and was the film's original director, was relieved of control of the project in October 2004. Lasseter put Brad Bird in charge of it nine months later. Bird inherited mostly finished sets, a group of characters, and a premise that appealed to him; he rewrote the script.

bank flipped television channels, all the images on the television were from Pixar commercials and shorts.

The film carried over one song from *Toy Story,* "You've Got a Friend in Me"; it was sung at different points by Tom Hanks and Robert Goulet. The film added two more Randy Newman compositions, Jessie's song—"When She Loved Me," sung by Sarah McLachlan over a montage animated by Tasha Wedeen—and the "Woody's Roundup" theme. Some audiences also heard a fourth Randy Newman song: Where American viewers saw Buzz Lightyear giving a speech in front of the Stars and Stripes with "The Star-Spangled Banner" playing, those outside the United States saw him giving it in front of a tableau of fireworks and a rotating globe, with an instrumental called "One World Anthem" supplying the score.

"When we went from a direct-to-video to a feature film and we had limited time in which to finish that feature film, the pressure really amped up," said Jackson. "Forget seeing your family, forget doing anything. Once we made that decision [on the schedule], it was like, 'Okay, you have a release date. You're *going* to make that release date. You're *going* to make these screenings.' "

All feature films at Pixar were difficult at the end; delivery dates loomed and hours inevitably became longer. Still, *Toy Story 2,* with its highly compressed production schedule, was an especially trying journey. While hard work and long hours are undoubtedly admirable up to a point—and Lasseter himself had put in his share of lonely overnights while working on his short films—running flat-out on *Toy Story 2* for month after month began to take a toll. For some staff, the overwork spun out into carpal tunnel syndrome, an occupational hazard of computer animation.

Pixar did not directly encourage the long hours and, in fact, set limits on how many hours employees could work by approving or disapproving overtime. An employee's self-imposed compulsion to excel, however, often trumped any other constraints. Jackson recalled that it was difficult to make young employees, especially, understand that "when I say you have two days, that doesn't mean I want you to spend two twenty-four-hour days to get this done."

For many, the clock was inconsequential compared to their desire to make a strong showing at their next director review—the session in which Lasseter or Unkrich would look at the project they were working on, be it artwork or a model or an animated shot. The situation came to a head when an overstressed and overtired animator set off to work with his infant child, having agreed with his wife that he would drop their baby at day care that morning. When he spoke with his wife later that day, she casually asked how the drop-off had gone—and he realized only then that he had, in his mental haze, completely forgotten. The baby was still in the backseat of his car in the parking lot. Although quick action by rescue workers headed off the worst, the incident became a horrible indicator that some on the crew were working too hard.

Yet the onerousness of a production was a subjective matter, which depended significantly on how one felt about the quality of the film.

"For animators, what's grueling is years and years of production with no end in sight, especially if you don't believe in the project," said Oftedal, who switched to *Toy Story 2* after *A Bug's Life.* "By comparison, *Toy Story 2* went by quickly. I believed the film was going to be good."

Unkrich regarded the film with pride while remembering the difficulty of meeting its due date. "Even though *Toy Story 2* really killed us in a lot of ways—it was really, really hard—I probably look back on that film the most fondly just in terms of how we all came together and did this impossible thing."

Pixar showed the finished *Toy Story 2* at CalArts on November 12, 1999, in recognition of the school's ties with Lasseter and more than forty other CalArts alumni who worked on the film; the students were captivated. The film had its official premiere the next day at the El Capitan Theatre in Los Angeles—the same venue as *Toy Story*'s—and released across the United States on November 24.

What had evolved from the twists and turns of *Toy Story 2*'s evolution was a work of surprising originality that delivered both emotional power and entertainment. Jessie's exuberance at meeting Woody, and later her anger toward him, flowed directly from her hard experiences; her effect on Woody, in turn, made it plausible for him to do the unthinkable by deciding he no longer belonged in Andy's room. The pensive side of the film coexisted easily, however, with its gentle comedy—mostly built around the toys' antagonists, toy collector Al McWhiggin, toy-store Buzz, and Emperor Zurg—and with its set-piece spectacles, such as the opening video game and the catastrophic street crossing.

Reviewers found the film to be that rara avis, a sequel that managed to equal or even outshine the original. " 'Toy Story 2' does what few sequels ever do," *The Hollywood Reporter* proclaimed. "Instead of essentially remaking an earlier film and deeming it a sequel, the creative team, led by director John Lasseter, delves deeper into their characters while retaining the fun spirit of the original film."

Variety held that *Toy Story 2* exceeded its predecessor as "a richer, more satisfying film in every respect."

> The only thing this sparkling picture lacks, by definition, is the shock of the new; four years ago, it was startling to behold the frontier that computer animation had conquered, to see what vast possibilities were now available in the whole field of animation. But there is no sense of complacency or sameness, as the filmmakers get their charges out of the house and into a situation that gives their lives more poignancy and awareness of mortality than, frankly, most characters in live-action films are accorded these days.

In all likelihood, even a mediocre sequel to *Toy Story* would have found an audience, such was the affection for *Toy Story*'s characters. Lasseter, however, was plainly hungry to surpass his own past achievements, and the come-from-behind artistic success of *Toy Story 2* was the result. The film was no less successful from a commercial perspective; it became the highest-grossing animated film of the year, earning $245 million domestically and $486 million worldwide—more than either *Toy Story* or *A Bug's Life* by a significant margin. It was the second-highest-grossing animated film of all time, behind Disney's *The Lion King.*

Pixar had beaten back DreamWorks for the moment, at least; Katzenberg's next computer-animated film, *Shrek,* would not release for another year and a half. Pixar had also established, if anyone was wondering, that the successes of its first two films were no flukes.

9. CRISIS IN MONSTROPOLIS

Monsters, Inc.

Lori Madrid had been writing songs for children since her days as a graduate student in social work in the mid-1990s. She continued her songwriting at her first jobs in treatment centers for abused children in Harlem and the South Bronx. Troubled children liked singing songs as part of their treatment, she found. Some of the songs were meant to help children feel better about themselves, or to help them get hold of their tempers. One song, "Strange House," helped children in foster care express what it was like to be taken away from their families and live somewhere else. Some of Madrid's colleagues at the centers started using her songs, as well.

Returning later to her native Torrington, Wyoming, she tried her hand at writing a children's story—a story in verse form that she titled "There's a Boy in My Closet." In twenty-eight lines, it told of a young monster who was frightened because he was sure he had seen a boy in his bedroom closet, while his mother insisted he had nothing to be afraid of—there was, she said, no such thing as boys.

In late October of 1999, Madrid mailed her story to around a half-dozen publishers, including Chronicle Books, a San Francisco–based house that published children's books along with books of art, architecture, and photography. She never heard back from Chronicle. One of the other publishers expressed interest in the story, but nothing came of it. Later, Madrid turned the story into a musical, *There's a Monster in My Closet,* which the local Big Brothers, Big Sisters program in Torrington produced and took around the state during the summer of 2001.

As the summer came to a close, several of her friends and coworkers who had seen the musical began urging her to see the trailer for an upcoming Pixar film, *Monsters, Inc.* They told her that the film was plainly based on her story. Pixar must have gotten hold of your work, they told her earnestly. Pixar stole from you.

Madrid saw the trailer herself during Labor Day weekend and reached the same conclusion. "I realized I was looking up at a movie screen and seeing my manuscript in living color," she recalled later. "Except that it didn't have my name on it. The closet, the child, the monster, the expressions, the scene—the total feel of the story was just like my creation."

She was distressed. Who would ever produce her musical now? Everyone would assume *she* had copied from the *film.*

Her musical was too recent for Pixar to have copied it, she thought. Someone at Pixar must have received her short story from one of the publishers. She did some research on the Internet and found that Pixar had just published a book with Chronicle, *The Art of Monsters, Inc.,* even though Pixar had published all its previous books with Disney's in-house publishing arm, Hyperion. What had happened was obvious, she concluded: Chronicle had passed her story to Pixar in 1999, and Pixar had reciprocated by switching to Chronicle.

She found a lawyer and filed suit in October against Chronicle Books, Pixar, and Disney in federal court in Cheyenne, Wyoming.

The allegations in the lawsuit seemed strained, at best. The idea of monsters who were afraid of children was just that, an idea, freely available for anyone to use. In fact, one of John Lasseter's own student films, *Nitemare,* dealt with a monster who was afraid of a little boy. Moreover, the ways of book publishing were such that an unsolicited submission from a novice author with no agent was unlikely to get so much as a glance from anyone higher than an editorial assistant, let alone be whisked into the hands of a major film studio.

Lastly, although Lori Madrid could not have known it, contacts between Chronicle and Pixar apparently did not start until August 2000, when production on *Monsters, Inc.* was well underway. A manager at Pixar was making inquiries of various publishers then about the idea of a Pixar imprint, or publishing brand, that would publish work from the studio's artists to keep them busy during downtime. That idea fell by the wayside, but Chronicle picked up the *Monsters, Inc.* art book in the spring of 2001 after Hyperion declined to publish it.

However unlikely the scenario that Madrid had laid out, film studios had to take such lawsuits seriously. If the case got as far as a jury trial, the defense lawyers could guarantee nothing. Pixar had once been part of Lucasfilm, and Chronicle published *Star Wars* books with Lucasfilm. Pixar and Chronicle were both in the Bay Area. It was within the realm of possibility for the jurors to conclude that somewhere in the web of relationships, someone at Chronicle had in fact handed the story to someone at Pixar in 1999, and that the story had made its way to the *Monsters, Inc.* team. The jurors could decide that the film infringed on Madrid's original expression in the story. Additionally, Madrid's lawyer, Beth Mary Bollinger, would have the chance to go through Pixar's files and e-mails, if she were so inclined—and there was always the chance that some stray comment might be turned against the studio.

Of more immediate concern, Bollinger had deployed the doomsday weapon of copyright litigation: She had asked the court to issue a preliminary injunction that would forbid Pixar and Disney from releasing the film while the suit was pending. The companies urged the judge to dismiss the suit immediately. Over their objections, however, the judge ordered a hearing on the motion for a preliminary injunction. It was to take place on November 1, 2001—the day before the scheduled release of *Monsters, Inc.* on some 5,800 screens in 3,200 theaters across the country.

Thus Pete Docter, the thirty-three-year-old director of the film, and Dick Cook, the fifty-one-year-old chairman of the Walt Disney Motion Pictures Group, found themselves in the federal courthouse in downtown Cheyenne on a windy Thursday morning. Joining them was a squadron of lawyers led by Steve Marenburg of the Los Angeles law firm of Irell & Manella, and by the defendants' local counsel, Terry Mackey. After Bollinger gave her opening argument and Madrid testified through the early afternoon, Mackey had his turn to approach the judge.

"Your Honor, the seriousness of this matter cannot be underestimated," he began. A preliminary injunction, he said, was "perhaps the most drastic and serious remedy" that the court could impose at this stage, and the consequences could be "disastrous." The plaintiff bore a heavy burden to justify such a step.

"We know this about the evidence," Mackey continued. "There has been no evidence of copying of the works of the plaintiff, none whatsoever. Indeed, there has been no evidence of access for copy of the work of the plaintiff by any—"

"There might be circumstantial evidence of that," the judge interjected.

"At best, Your Honor," Mackey replied. He argued that it was "a long reach."

Mackey then turned to the issue of whether Madrid's story had been infringed at all. "The Court must look at the works side by side and lay them down because you can only compare the words," he said. The concepts that Bollinger pointed to—such as a monster's fear of a child, and the closet as a portal between the human world and a monster world—were not original to Madrid's story, he added.

Docter took the stand a few minutes after 3 p.m., with Mackey examining him.

Q: Would you state your full name, please.
A: My name is Peter H. Docter.
Q: Spell your last name, please.
A: D-O-C-T-E-R.
Q: What's your occupation?
A: I'm a director at Pixar Animation Studios.
Q: When did you become a director at Pixar Animation Studios?
A: Roughly in 1997.
Q: And can you tell the Court, please, what role you played in the original development of a movie for Pixar about monsters?
A: Yes, sir. As the director I was involved in everything in the film from concept to story to voice casting and direction to character design to final completion.

Mackey led him through the progression of the story's development. Docter said the story lines were original, uninfluenced by Madrid's story or any other source—"only our own childhoods," he said.

The presiding judge was Clarence Brimmer, Jr., a genteel seventy-nine-year-old former attorney general of the state, appointed to the bench during the Ford administration. After Docter referred to the casting of John Goodman and Billy Crystal as the voices of the lead monsters, Judge Brimmer interrupted to ask, "One of my deputy clerks told me that those names are the names of people who are prominent actors. Is that correct?"

Mackey affirmed that the judge had been told correctly.

Bollinger focused her cross-examination on an issue of whether Boo—the little girl who crosses through her closet door into the monster city, Monstropolis—had really been unchanged since 1999. A description from early 1999 had her "often more monstrous than her monster captors," with a "bold, stubborn nature" and a "tough, survivor instinct." In Bollinger's view, the Boo character in the film—or, at least, in the clip that Mackey showed—was far different.

Q: So you're describing that little girl as having a tough survivor instinct?

A: Somewhat. It is not especially evident in the clip we just saw.

Q: And is it evident in the movie?

A: She's pretty fearless and scrappy.

Q: What makes her fearless and scrappy? Can you describe a scene that makes her fearless and scrappy?

A: The fact she's come face-to-face with an eight foot tall, 2,000 pound, big, blue, furry guy and says, "Duck it," to him fairly describes that. There's also sequences in the film where she runs around the apartment completely oblivious to the two monsters.

As Docter wound up his hour on the witness stand, Judge Brimmer offered some help. "It [the film] has taken you, what, five years of hard work and, I take it, a massive amount of money?"

"Yes," Docter answered, "Not mine, but—"

"And it would take the work of many, many people?"

"That's correct, yes. Over 450, I believe, worked on the film. And yes, in this case the production from my initial concept to completion was five years, which is fairly typical. *Toy Story* took about four and a half years."

"So this is nothing that could be done overnight," Judge Brimmer continued, "and it wouldn't be anything that you would do on a whim after you had read somebody else's poem?"

"No, no, definitely not," Docter agreed.

Marenberg, Disney's and Pixar's counsel from Los Angeles, put Dick Cook on the stand. Cook had started at Disney three decades earlier as a ride operator running the steam train and the monorail, then moved steadily upward through the ranks. Now he was in charge of the production, distribution, and marketing of all Walt Disney Co. films and videos around the world. Marenberg questioned him about the effect of a preliminary injunction against the release of the film.

The effect on Disney would be devastating, Cook said. *Monsters,*

Inc. was one of the industry's "tent pole" films for the season, along with *Harry Potter and the Sorcerer's Stone* and *Pearl Harbor.* Disney had set the date far in advance, close to a year ago. Disney had primed audiences with about forty thousand trailers in movie theaters and a costly ad campaign. There had been a "giant press junket" two weeks earlier with Docter and Lasseter and the film's stars. The company had already spent about $3.5 million on a premiere and special screenings. Everything had been choreographed to peak on November 2. Tomorrow.

The 5,800-odd prints, he said, had already gone out from Technicolor's warehouses in California and Ohio and were sitting at theaters. Theater owners who were counting on showing the film would have dark screens.

"The reason we chose November second was because there's a lot of competition coming later on at Thanksgiving with *Harry Potter* and at Christmas with some other pictures," Cook explained. "And what we wanted to do, because of the importance of this movie, is to be the first one out."

Q: So obviously the delay of the film by injunction or otherwise would affect the first weekend and the ability to gain all of the benefits you've gotten by virtue of the fact that November second is the first weekend?

A: It would be a disaster.

Q: And that would affect, then, not only the theatrical performance of the film, but what other markets in the United States?

A: Well, it would completely be a snowball effect in a reverse way in that it would certainly put a damper on all of the home video activities, all the DVD activities; in fact, would influence international because international is greatly influenced on how well it does in the United States, and by taking that away, it would definitely, definitely, have a big, big impact on the success of the film.

And furthermore, going further, is that it would take away any of the other ancillary things that happen, you

> know, whether it would become a television series, whether or not it becomes a piece of an attraction at the parks, whether it becomes a land at the parks, or any of those kinds of things.

The life cycle of a film, Cook explained, was like that of fruits and vegetables. "You try to plan it well in advance and you lay your seeds, put your fertilizer on, you water it, make sure it is nourished and all of that. When it is ready to pick, it is ready to pick."

Pixar's film was ready to pick. Cook told Judge Brimmer that Disney had hired the National Research Group, a firm that carries out polling for a number of studios, to gauge awareness of *Monsters, Inc.* "It is just off the map," Cook said of the results. "In our terminology, it is in the 80's and normal movies are in the 30's or 40's."

"*Monsters, Inc.* is in a sense the fruit that's ripe today?" Marenberg asked.

"It is the peach that you're holding that's ripe today and you want to eat it now," Cook said. "If you don't eat it now, it's going to spoil."

Marenberg turned to Judge Brimmer. "Nothing further," he announced.

After hearing final arguments from Bollinger and Mackey—Bollinger waived cross-examination of Dick Cook—Judge Brimmer ruled from the bench. He denied the defendants' motion to dismiss the lawsuit and then went through the four requirements for a preliminary injunction: Would Madrid suffer irreparable harm without the injunction? Was she substantially likely to win the case in the end? Would the harm to her from the release of the film outweigh the harm to Disney and Pixar from the injunction? Finally, would the injunction be contrary to the public interest? As he ticked off each issue, he explained why he felt the balance favored letting the film open on time.

Reaching the last of the issues, that of the public interest, he offered half facetiously, "My law clerk pointed out that he has a couple of nieces, two and seven, that would be very upset if I issued this injunction. . . . I suspect that there are a lot of little kids all over the country that would regard me as the worst kind of judicial monster if I were to do it."

The doomsday weapon had been defused. Disney and Pixar would be able to open *Monsters, Inc.* Meanwhile, the case would proceed, and if Madrid were able to show copyright infringement, the companies would face an award of damages.

The parties to a civil suit enjoy the right to "discovery," that is, to demand that the other side respond to written questions under oath, make its employees and other witnesses available for depositions, and turn over documents. During the ensuing months, Bollinger deposed Lasseter and *Monsters, Inc.* production co-designer Bob Pauley. Pixar archivist Christine Freeman gathered more than a hundred boxes of materials from the company's files in response to Bollinger's demand for, among other things, "all writings in your possession that relate to the storyline of *Monsters, Inc.*" and "all art, including concept art, related to *Monsters, Inc.*"

Judge Brimmer brought the machinery to a halt on June 26, 2002, when he ruled—in response to a motion from the defendants—that the film simply had nothing in common with Lori Madrid's twenty-eight-line poem apart from some general themes and ideas that copyright law did not protect. "All of these ideas are standard and indispensable with these characters, and with children's stories in general," he wrote. "To say that such things may be protected by copyright laws would . . . certainly be chilling to the free flow of children's stories, and could mean that many a child's bad dream would be a copyright infringement."

The *Madrid* case was now safely in the past. Pixar had endured the litigation process and won. As events would unfold, however, the case would prove to be merely a prelude to another, more problematic lawsuit still to come.

Pete Docter began work on the film that would become *Monsters, Inc.* in 1996 while others focused on *A Bug's Life* and *Toy Story 2*. Its code name was *Hidden City,* so named for Docter's favorite restaurant in Point Richmond. The protagonist initially was to be a thirty-two-year-old man coping with monsters that only he could see. After Docter scrapped this concept, the lead human character became a little boy for a while, and ultimately a little girl.

By early February 1997, Docter had drafted a treatment together with Harley Jessup, Jill Culton, and Jeff Pidgeon that bore some resemblance to the final film. In that story, titled simply *Monsters,* the character of Sulley (known at this stage as Johnson) was an up-and-comer at his workplace, where his job was to scare children; his eventual sidekick, Mike Wazowski, had not yet been added.

> A kid goes into his basement to find his missing sock. From out of a pile of laundry, a fearsome monster rises up and scares the bejeezus out of the kid.
>
> The monster, a big hairy blue guy named Johnson, emerges from a door. He removes his false fangs, pointy nose, and false humps. He fills out his scare paperwork, which he signs in triplicate. . . .
>
> Johnson walks from the workroom into a lounge area and delivers his paperwork to the dispatch cage. The dispatcher pulls a lever which sends the door off to the Door Storage Area. Johnson sits down next to his fellow monsters to wait for new assignments. Unlike his co-workers who play cards, smoke, or chat among themselves, Johnson keeps busy reading up on the latest scare technique. Much like Jack Lemmon in "The Apartment," Johnson is an attentive, tense enthusiast who lives for his work.

Things go wrong for Johnson during his next scaring assignment, when he accidentally brings a six-year-old girl back to the locker room with him. (The girl, called Mary at this point in the process, will later take the name Boo at Andrew Stanton's suggestion.) Mary screams when she first sees Johnson, but later calms down and demands to be taken home. Johnson struggles to get her there without attracting notice; children are forbidden in the monster world.

On February 4, Docter pitched the story to Disney along with some initial artwork, describing *Monsters* as a buddy story between

Johnson and Mary. He and his story team left with some suggestions in hand and returned to pitch a refined version of the story on May 30. At this pitch meeting, a longtime Disney animator named Joe Grant—he went all the way back to *Snow White and the Seven Dwarfs,* Disney's first animated film—suggested the title *Monsters, Inc.* The name stuck.

In a subsequent treatment, on August 8, Mary became a fearless seven-year-old who had been toughened by years of teasing and pranks from four older brothers. Johnson, in contrast, is nervous—nervous about the possibility of losing his job as a scarer after the boss at Monsters, Inc., announces that a downsizing is on the way. He feels envious of another scarer, Ned, the top performer in the company. "He believes that if he could be as good a scarer as Ned, the rest of his life would fall into place and he wouldn't be so unhappy."

The idea of a monster buddy for the lead monster emerged at an April 6, 1998, story summit in Burbank with Disney and Pixar employees. Such a character, the group agreed, would give the lead monster someone to talk with about his predicament. Docter named the character Mike for the father of his friend Frank Oz, the director and Muppet performer.

Like Woody and Buzz before them, Sulley and Boo went through radical changes as the story evolved between 1996 and 2000. Sulley went from a janitor, to a refinery worker, to a former scarer working in a refinery because an accident cost him his eyesight, to his final incarnation as the best scarer at Monsters, Inc. His body gained and lost tentacles as well as a spiderlike configuration of multiple eyes. After Boo started out as a grown man and then turned into a little boy, the character evolved into a domineering, out-of-control little girl—akin to the kidnapped boy in O. Henry's story "The Ransom of Red Chief"—before becoming the mild, innocent, preverbal girl of the film. At one point after Boo became a girl, she was to be from Ireland, mystifying Sulley and Mike by repeatedly referring to Mike as a "wee leppy karn"; at another point, she was to be Pixar's first African American character.

The voice role of James P. "Sulley" Sullivan went to John Goodman, the longtime co-star of the comedy series *Roseanne* and a regular in the films of the Coen brothers. Goodman interpreted the character to himself as the monster equivalent of a National Football League player. "He's like a seasoned lineman in the tenth year of his career," he said at the time. "He is totally dedicated and a total pro."

Billy Crystal, having regretted turning down the part of Buzz Lightyear, accepted that of Mike Wazowski, Sulley's one-eyed best friend and scare assistant. Boo was voiced by the preschool daughter of one of the story artists.

In November 2000, early in the production of *Monsters, Inc.,* Pixar picked up and moved for the second time since its Lucasfilm years. The jokes about Pixar's quarters in Point Richmond—"just turn left at the refinery"—had been getting worn. The company's approximately five hundred employees had become spread among three buildings, separated by a busy highway. At the going-away party marking the departure from the Point Richmond offices, someone ritualistically smashed a computer with a sledgehammer in a moment reminiscent of the then-recent film *Office Space.*

For the new facility, Jobs had initially considered former warehouse space in San Francisco, but it would have been inconvenient both to him and to the many Pixar employees who had settled in the East Bay and in Marin and Sonoma Counties to the north. He finally opted for a location in downtown Emeryville, California (population 6,900), near Berkeley, where he was able to secure a former Del Monte cannery site on sixteen acres.

Lasseter at first pressed for a campus in the pattern of a traditional Hollywood lot, with separate buildings for the different films in production and bungalows for development; he pulled back from that idea after a conversation with his mentor Frank Thomas from Disney. Thomas bemoaned the day when Walt moved the animators from a crowded building on Hyperion Avenue in Los Angeles to a spacious, seemingly paradisical complex in Burbank with private offices and individual buildings for each department. It turned out that everyone lost touch, Thomas said. With that, Jobs and Lasseter reconceived Pixar's facility as a single large building—it would be some 218,000 square feet—laid out to encourage spontaneous encounters in its long atrium. (It was designed, as it happened, by the firm that co-designed Bill Gates's massive lakefront residence near Seattle.) Employees quickly dotted the interior with whimsically themed workspaces; a tiki hut facade stood next to that of a suburban-style house with a flag, mailbox, and satellite dish.

In production, *Monsters, Inc.* differed from earlier Pixar features in that each main character had its own lead animator: John Kahrs on Sulley, Andrew Gordon on Mike, and Dave DeVan on Boo. Kahrs found that the bearlike quality of Goodman's voice provided an exceptionally good fit with the character. He faced a difficult challenge, however, in dealing with Sulley's sheer mass; traditionally, animators conveyed a figure's heaviness by giving it slower, more belabored movement, but Kahrs was concerned that such an approach to a central character would give the film a sluggish feel. Like Goodman, Kahrs came to think of Sulley as a football player, one whose athleticism enabled him to move quickly in spite of his size. To help the animators with Sulley and the other large monsters, Pixar arranged for Rodger Kram, an expert at Berkeley on the locomotion of heavy mammals, to come in and lecture on the subject.

Adding to Sulley's lifelike appearance was an intense effort by the technical team to refine the rendering of fur. Other production houses had tackled realistic fur, most notably Rhythm & Hues in its 1993 polar bear commercials for Coca-Cola and in its talking animals' faces in *Babe. Monsters, Inc.,* however, required fur on a far larger scale. From the standpoint of Pixar's engineers, the quest for fur posed several significant challenges. One was figuring out how to render the huge numbers of hairs—2,320,413 on Sulley—in a reasonably efficient way. Another was making the hairs cast shadows on other hairs. Without self-shadowing, fur or hair takes on an unrealistic flat-colored look. (The hair on Andy's toddler sister, as seen in the opening sequence of *Toy Story,* is an example of hair without self-shadowing.) Still another was giving the animators control over the direction and movement of the fur without imposing an inordinate burden on them.

"For anything with complex physical movements—hair, cloth, fluids, etc.—it's impractical for a person to hand-animate every detail," observed Tom Lokovic, who worked on the fur team. "You want things to behave 'physically,' which means you run a simulation that drives the hair or cloth or whatever according to some physical model."

> A lot of work goes into choosing the parameters to make a simulation look right. Set one parameter too low, and the hair droops like it's made of rubber bands; set it too high, and the numerics go unstable, causing the hair to "explode."
>
> As important as simulation is, you can't always let physics do all the work. . . . For example, for dramatic reasons you might want the hair to end up in a certain place on a character's face, even if "physics" wouldn't have put it there. In a case like that, you can introduce additional physical effects, such as invisible collision objects or extra forces, to get the simulation to do what you want, or just override the results of the simulation, telling the geometry explicitly where you want it to go. In practice, it's a combination of both.

The fur simulation techniques became part of a new program called Fizt (for "physics tool"). After a shot with Sulley had been animated, Fizt took the data for the shot and added his fur, taking into account his movements as well as the effects of wind and gravity.

The Fizt program also controlled the complex folding and wrinkling movement of Boo's loose-fitting T-shirt. In earlier Pixar features, the clothing on human characters stayed unnaturally smooth and did not move independently of the character's body; in some cases, clothes were essentially painted textures on the character's skin. The nucleus of the technology that Fizt employed to make

Boo's T-shirt act like a shirt originated with the cloth simulator that senior scientist Michael Kass created for Geri's jacket in *Geri's Game.*

The deceptively simple-sounding task of animating cloth meant solving the complex problem of how to keep cloth untangled—that is, how to keep it from passing through itself when parts of it intersect (for instance, when a character pinches its clothes by bending its shoulders, elbows, or knees). Kass, joined on *Monsters, Inc.* by David Baraff and Andrew Witkin, developed an algorithm they called "global intersection analysis" to handle these cloth-to-cloth collisions.

The complexity of the shots in *Monsters, Inc.*—including not only Sulley's hair and Boo's shirt, but also elaborate sets such as the door vault with its racks of a half-million hanging doors—required more computing power to render than any of Pixar's earlier efforts. The render farm in place for *Monsters, Inc.* was made up of thirty-five hundred Sun Microsystems processors, compared with fourteen hundred for *Toy Story 2* and two hundred for *Toy Story.* (Each computer in the render farm generally had two or four processors running in tandem, so there were far fewer computers than processors.) Indeed, when Pixar studied the total amount of computational power needed on each film, factoring in that the newer films were rendered not only on more processors, but also on newer, faster ones, the results showed that *Monsters, Inc.* required more power than Pixar's three previous feature films put together.

Released on November 2, 2001, to supportive reviews, the film went on to take the place of *Toy Story 2* as the second-highest-grossing animated film of all time, behind only *The Lion King.*

In a transparent effort to dilute interest in *Monsters, Inc.,* DreamWorks continued its rivalry by releasing the home video DVD of its recent hit *Shrek* the same day *Monsters, Inc.* opened in theaters. It was a testimony to the growing power of home video and to the popularity of the DreamWorks film that the *Shrek* DVD grossed more that weekend in retail sales than *Monsters, Inc.* did at the box office. Jeffrey Katzenberg could take further satisfaction the following March when *Shrek,* not *Monsters, Inc.,* won the first Academy Award in the newly instituted category of best animated feature—possibly because the more adult, cynical humor of *Shrek* held greater appeal for Academy voters.

On October 1, 2002, not quite eleven months after the release of *Monsters, Inc.,* a second copyright infringement suit materialized. The defendants, once again, were Disney, Pixar, and Chronicle Books. The plaintiff, Stanley Miller, was a sixty-two-year-old professional artist and illustrator, best known for his concert posters and album covers for the Grateful Dead. He invariably signed his work with his adopted professional name, Stanley Mouse.

As a teenager in the late 1950s, Miller had started his artistic career drawing monster hot-rod cartoons, and he continued drawing and selling cartoons for posters, T-shirts, and calendars. Two of his stock characters since the early 1960s were a large, hairy monster and his smaller, one-eyed companion, both of whom went through various stages of evolution over the decades. In 1997, Miller prepared sample artwork, a treatment, and a script for an animated film titled *Excuse My Dust,* in which the large monster would appear under the name Fred Flypogger and the smaller one would be named Wise G'Eye. They would reside in a place called Monster City, which was operated by Monster Corporation of America (also the name of Miller's real-life company). Miller circulated the materials for *Excuse My Dust* the following year.

To the casual onlooker, Sulley and Fred Flypogger seemed to have little in common except that they were both bipedal, hairy monsters. The designs of Mike and Wise G'Eye, however, *did* bear an intriguing resemblance, both of them having heads that consisted of a large eyeball and mouth perched on two spindly legs. Also, Wise G'Eye had a taller girlfriend, Lucretia, whose waist was in line with the top of his head, much like Mike with his girlfriend Celia. During the lawsuit's discovery process, moreover, it emerged that Pixar's art department had gathered samples of Miller's cartoon art—though there was no proof the collection included Fred Flypogger or

Wise G'Eye. It also developed that Mike had been designed *after* Miller circulated *Excuse My Dust,* and that one of the recipients of the artwork was a friend of Lasseter's. (The friend stated in a deposition that he had not shown Lasseter the material.)

In one sense, the contentions in the lawsuit were a statement of the obvious: Anyone making an animated film about monsters would likely be influenced by other artists whose work they admired and who had created interesting depictions of monsters. In art and drama, as in other endeavors, rarely was there anything completely new under the sun. Shakespeare famously drew from the Greek writer Plutarch for his Roman tragedies.

At Pixar in particular, looking at the approaches of other artists was part of the culture that Lasseter had instilled from the beginning. Both before and after *Monsters, Inc.,* Pixar's films showed clear cinematic and literary influences—some large, some small—to which few would object. Lasseter often spoke of Hayao Miyazaki's influence on his work. The *Toy Story* team drew inspiration from popular American buddy films. *Toy Story* shared its fundamental premise (that toys come alive when no one is looking) and some plot points (a toy that doesn't know it's a toy arrives as a birthday present, goes missing, and must be rescued) with 1977's *Raggedy Ann and Andy,* a film directed by Lasseter's onetime boss Richard Williams.

A Bug's Life was a wry comic twist on the classic western *The Magnificent Seven* and its Japanese predecessor, *The Seven Samurai*—in much the same vein as the 1986 John Landis film *¡Three Amigos!* The plot of *Cars* followed that of the 1991 Michael J. Fox feature *Doc Hollywood,* while the character designs in *Cars* reflected the influence of Bill Peet's car designs for the 1952 Disney short *Susie, the Little Blue Coupe.* A pivotal moment of *Ratatouille*—a restaurant critic being carried away by a lowly peasant food—harked comedically to the madeleine incident in Marcel Proust's *Remembrance of Things Past.* Alfred Hitchcock's *Rear Window* was imprinted on two Pixar films, eight years apart: in the camera-flash-as-weapon scene of *Toy Story 2* and in the apartment vignettes witnessed by the star of *Ratatouille.*

Art was never created in a vacuum chamber. The difficult question posed by *Miller v. Pixar Animation Studios* was, if *Monsters, Inc.* had drawn upon Miller's characters, when did homage and artistic influence cross the line into misappropriation?

Both Miller and the defendants hired expert witnesses and submitted their reports to the court in August 2004. One of the defense experts, Jeffrey Cohen, was an English professor at George Washington University in Washington, D.C., and an authority on the history of "monstrosity" in literature and culture. He had written several books on the subject, namely *Monster Theory: Reading Culture; Stories of Blood: Monsters, Jews and Race in Medieval Britain;* and *Of Giants: Sex, Monsters, and the Middle Ages.* Cohen explained that one-eyed creatures were long-standing residents of literary monsterdom. The Freemason Eye of God on the American dollar bill, the ethereal eye of the evil sorcerer Sauron in the novel *The Lord of the Rings,* the Cyclops of Greek mythology, the alien Leela from the animated television series *Futurama*—all were precedents for one-eyed creatures such as Mike or Wise G'Eye. "Miller was working in a long established and rich tradition," Cohen wrote.

Cohen added that the pairing of male characters with dissimilar body types as buddies could be found in the John Hughes film *Planes, Trains & Automobiles* with Steve Martin and John Candy, the novel *Of Mice and Men,* and elsewhere.

Stanley Miller was not claiming, however, to have invented the one-eyed monster or the odd-couple story. With regard to the question of whether Miller's character Wise G'Eye appeared to have influenced Pixar's character Mike, Cohen merely stated his conclusion that they "are not substantially similar representations of monsters" without elaborating.

The defendants' other expert, Stephen Bissette, was a comic-book writer and illustrator, a commentator on horror and fantasy films, and co-manager of an award-winning video store in Brattleboro, Vermont. Like Cohen, he offered an impressive study of cyclopean figures throughout history and then abruptly concluded that Wise G'Eye and Mike "are unique from one another in substance and in those distinctive particulars which set them apart from one another and the generic archetype."

Both Cohen and Bissette further noted that Miller's hairy-man figure was likewise well precedented, including in stories of Bigfoot and the Yeti.

Miller's expert witness, Jerry Lee Brice, took a different tack, focusing on the evolution of the lead characters in *Monsters, Inc.* and on how Mike and Wise G'Eye compared to each other. Brice was a CalArts animation graduate who had worked for Disney Feature Animation and other animation studios and was a member of the animation faculty at the Art Institute of California at San Diego. In his report, he listed ten *Monsters, Inc.* artists with whom he had worked in the course of his career, including Docter. "I have great respect for each of those artists and my time working with them was very positive," he added.

His conclusions were not so congenial. Brice wrote that after reviewing the documents and artwork Pixar had turned over during the discovery process, he believed "the character of Mike Wazowski is so similar to the Stanley Miller creation Wiseguy from the *Excuse My Dust* television series and motion picture presentation pitch that . . . the design and personality must not only have been inspired by, but must also have been copied, directly from Mr. Miller's original creation."

Mike, he wrote, emerged in a manner "profoundly different than the development track for the other characters by the same artists in the same movie." Where Sulley and Boo went through a typical design process of numerous stages, Mike seemed to materialize whole-panoplied on a storyboard, fully realized from the beginning. Brice also found it noteworthy that in early sketches of Mike, the character was drawn "in the exact same way as Mr. Miller's pen and ink version of the character on artwork dating back to 1963."

> I have never seen two artists sketch exactly the same way in my life. Sketching is like handwriting. Every individual has a

> unique signature and style to sketching and drawing letters. The only time such identical similarity happens is when an artist has mimicked and practiced another artist's style and consciously tries to copy someone else's artwork.

Brice also argued that pervasive similarities between the character designs pointed to copying. Apart from the obvious parallels—the same round head-and-torso combination, the same placement of the mouth and eye, the same shade of green skin—Brice noted that both characters had a fold of skin where an eyebrow would normally be; shoulders growing out of the area where the ears would normally be; and a jaw hinge located directly in front of their shoulders. Mike and Wise G'Eye both had slender arms of the same unusual length, nearly reaching their feet.

Disney, Pixar, and Chronicle asked the judge to grant "summary judgment"—to determine that they were legally entitled to win on the grounds that Miller had not secured copyrights in his early drawings and that there was no evidence Pixar had received a copy of *Excuse My Dust.* On May 2, 2005, the judge assigned to the case, Judge James Ware of the U.S. District Court for the Northern District of California, handed down his decision. He agreed with the defendants on the first issue: Miller's drawings from the early 1960s and for the next forty years were without copyright protection. Miller, it seemed, was a better artist than he was a businessman; he had consistently either failed to copyright his works containing the predecessors of Fred Flypogger and Wise G'Eye during that time or failed to renew the copyrights when they came up for renewal. As a result, only *Excuse My Dust* was copyrighted by the time Pixar made *Monsters, Inc.*

On the second issue, however, Ware ruled in Miller's favor. It was possible, Ware said, that Lasseter's friend (an artist at Industrial Light & Magic) had shared a drawing of Wise G'Eye from *Excuse My Dust.* Whether he had actually done so was a disputed factual issue; hence, it would to be up to a jury to decide whom to believe. Moreover, even if the jury found no direct evidence of Pixar's access to

Excuse My Dust, it could still infer that Pixar had copied the character if it found "striking similarity" between Wise G'Eye and Mike. Ware said he saw enough similarities between the two characters that the case could proceed to a trial.

The case ended inconclusively. In January 2006, before the case went to trial, Pixar and the other defendants entered into a settlement with Miller on undisclosed terms. The Pixar artwork and files turned over during the discovery process remained under seal, as did the depositions.

Shortly afterward, on a blog that Professor Cohen operated with other scholars, the defense expert posted some circumspect reflections on the case beneath an image of a wide-eyed Mike Wazowski:

> It seems that a famous illustrator known for his depictions of hot rods and psychedelia thought that Mike bore a little too much resemblance to one of his own animated eyeballs. The case just settled, and legally I can't talk about it much, but suffice it to say that sometimes a medievalist can be useful to corporate America. As an expert witness I was hired to research and compose a report on one eyed monsters throughout human history. Frankly I was surprised at how many I uncovered—proof, I think, that the human imagination has always been haunted by body parts endowed with an unnerving autonomy. A bigger claim could even be advanced that central to the monstrous is the body in pieces, the flesh that isn't governed by a unifying soul but keeps exerting its unpredictable will.
>
> It was fun, it was a glimpse into a world where half a billion dollars could be at stake, I got to give a deposition and be grilled about lime green versus avocado green skin and its signification, but now it is over.
>
> It reminded me quite forcefully, too, that monsters never seem to cede their power to haunt.

10. EMERYVILLE

Finding Nemo, The Incredibles

In 1992, in the midst of story work on *Toy Story,* Andrew Stanton visited what was then an oceanarium and wildlife park at nearby Marine World. There, he found the starting point for what would be Pixar's fifth feature film, and the second one that he would co-direct. Staring into the tanks, he became intrigued by the idea of reproducing the world of sea creatures in computer animation. He also found inspiration in reflecting on his relationship with his young son.

"When my son was five, I remember taking him to the park," he said.

> I had been working long hours and felt guilty about not spending enough time with him . . . but I spent the whole walk going, "Don't touch this. Don't do that. You're gonna fall in there." And there was this third-party voice in my head saying, "You're completely wasting the moment you've got with your son right now." I became obsessed with this premise that fear can deny a good father from being one.

During production of *A Bug's Life,* Stanton found time to draft a version of the story and presented a pitch to Lasseter for an hour. Afterward, an exhausted Stanton waited for Lasseter's reaction. Lasseter, a longtime scuba diver, told him, "You had me at 'fish.' "

The story told in *Finding Nemo* was that of a father clown fish, Marlin, who had lost his wife and all but one of his children in a barracuda attack. He was thus desperately overprotective of his surviving son, Nemo—only to see Nemo, on his first day of school, netted by a scuba diver and taken away to parts unknown. The film then traced Marlin's odyssey in search of Nemo, aided by a forgetful blue tang named Dory, while Nemo sought to escape from the tank in the dentist's office where he had ended up.

The evolution of the script proved to be an illustration of how releasing information to the audience, or withholding it, could drastically change the tone of a film. The script originally gave the story of Marlin's first tragedy, the loss of his family, in small pieces interspersed with his journey. Early in the film, audiences were to see a flashback to Marlin's first sighting of his wife and how they met. Later, the film would flash back to Marlin and his wife moving into their new home, a safe haven of anemone tentacles. Still later, a flashback would show Marlin as a father-to-be helping his wife prepare to give birth. Another would show the two parents watching contentedly over their hundreds of eggs. Then, *finally,* in the third act of the three-act film, during the film's fishing-boat sequence, the audience would learn of the barracuda attack that had devastated Marlin and left him so overprotective.

Far along in the story process, Stanton became dissatisfied with the flashback structure. It was too convoluted, he felt. Moreover, he believed audiences wouldn't like Marlin because they wouldn't understand his fearful personality until near the end of the film; Marlin's protectiveness would seem merely annoying. Stanton reduced the five flashbacks to one short sequence at the beginning of the film, one that related (but did not directly show) the barracuda attack.

"Boom, you suddenly cared about Marlin," Stanton said. "I didn't have to change any lines. I didn't have to change any readings. He suddenly wasn't annoying anymore. He was somebody you empathized with."

The part of Marlin was originally to be voiced by the actor William H. Macy, who was most recently a co-star of the film *Magnolia.* After Macy had recorded his dialogue, however, Stanton decided the part needed a lighter touch, and the comic actor Albert Brooks took his place. Stanton wrote the part of Dory, Marlin's sidekick, with Ellen DeGeneres in mind. While Stanton's wife was watching DeGeneres's television show *Ellen* one evening, he had happened to see her character "doing her schtick of changing her mind five times before a sentence finishes," he recalled. From that moment onward, Dory and DeGeneres were inseparable in his mind. (When he later phoned her to see whether she would consider the part, he explained that he'd written it for her and would be sunk if she didn't take it. She told him, "Well, then, I guess I'd better take it.")

Dory's character was more than comic relief; as Stanton saw it, her memory loss made her innocent like a child—a substitute child for Marlin during his quest. Dealing with her would force him to learn a modicum of patience and tolerance for her venturesome risk taking, preparing him to be a better father when he finds his son.

Stanton also spoke of a spiritual aspect to the relationship of Marlin and Dory. Dory was, literally, an angel fish. "The protagonist's battle was to overcome fear by discovering faith, and certainly Dory represented the angel, or the helper who showed him how to let go and not be consumed by his worries," he told an interviewer for a Christian-oriented film Web site.

He observed that subtlety is critical in giving films such as Pixar's a spiritual or religious dimension. "My personal view is that if you go into things on a pulpit or with an agenda in the creative world, it can easily get in the way of your creativity and quality. . . . Be Christ-like in everything you do, not worrying about whether you're furthering the cause."

The underwater setting of *Finding Nemo* had no counterpart in any previous Pixar film. The animation of walking and gesturing, the interplay of light and shadows in a room or the outdoors—a host of familiar concepts would have to be jettisoned or reconceived. For inspiration, team members screened underwater scenes from Disney films such as *Pinocchio, The Sword in the Stone,* and *The Little Mermaid.*

Stanton and executive producer Lasseter finally embraced a naturalistic approach. To give artists, animators, and engineers a frame of reference, they screened Jacques Cousteau documentaries and the IMAX film *Blue Planet,* as well as *Jaws* and *The Abyss.* They had a twenty-five-gallon fish tank brought in and stocked with an assortment of saltwater fish. At Lasseter's direction, some team members traveled to Hawaii to scuba dive at company expense and understand better what the environment was like down below.

The studio's in-house training department, Pixar University, came into play. Normally, the department offered general artistic and technical courses to help employees improve their skills and broaden their horizons and it administered the studio's archives. At times, however, it also responded to factual questions in the course of a film's development. Early in the development of *Finding Nemo,* Elyse Klaidman, a Pixar University manager, was assigned to bring in a fish expert to give a lecture to the project's still-minuscule team.

As it happened, a tenant in her house, Adam Summers, was a postdoctoral fellow at Berkeley with an encyclopedic knowledge of the subject. She invited him to Pixar to talk for an hour. When he asked her why they were interested, Klaidman answered simply, "I can't tell you." (There would be no repetitions of the *Antz* vs. *Bugs* situation if Pixar could help it.) He came and spoke in front of seven or eight people, Stanton among them. Two and a half hours later, he left, feeling exhilarated—not because he was a Pixar fan (he had never seen an animated film), but because he was amazed at the intensity of their interest.

"I had never encountered students anything like this," Summers remembered. "This was like the best graduate student class I'd ever taught. I couldn't get more than three or four minutes of talking in before someone would raise their hand and ask questions that would send us off in all sorts of different directions."

A couple of weeks later, Klaidman asked him to come back on a regular basis. The company had him sign a nondisclosure agreement, then explained what the film was and showed him some test footage. He became the lead aquatic consultant, returning frequently to give advice on the appearance and behavior of the fish species in the film.

"It felt like people genuinely wanted to hear my complaints and my poking holes in what they were doing," he recalled. "If a fish was doing the wrong thing, they actually wanted to know."

At times, the filmmakers departed from technical accuracy, but only to serve the purposes of the story. At first, Dory swam without wiggling her tail—correct for her species, which relies only on its fins to move, but Lasseter objected because her movement would not look plausible to an audience. The Dory of the film wiggles her tail. Summers explained that male clown fish change sexes when the dominant female dies; nonetheless, Marlin does not change sexes in the film. On one occasion when Summers objected strenuously to a depiction of fish behavior, Ricky Nierva, the art director for character design, reminded him, "Adam . . . fish don't talk."

Real fish also lack expressive faces. Such expressiveness was crucial in *Finding Nemo* because the fish characters had no arms or shoulders with which to gesture. To enable the fish to act and emote, the designers added a few human touches to their faces while retaining an essentially fishlike appearance: They brought the fish's eyes to the front, added eyebrowlike accents, and gave them flexible lips.

Overall, though, the studio's commitment to realism was fanatical. To prepare for the scene in which Marlin and Dory become trapped in a whale, two members of the art department climbed inside a dead gray whale that had been stranded north of Marin. Crew members dissected dead fish to learn their anatomy—the muscle, the heart, the gills, the swim bladders. Summers brought a series of world authorities to Pixar, including Mark Denny of Stanford, who spoke on waves; Terrie Williams of the University of California at Santa Cruz, a whale expert; Matt McHenry of Berkeley, who explained jellyfish propulsion; and Mimi Koehl of Berkeley, who detailed the movement of algae and sea grass. Because many objects in the sea are translucent—neither fully transparent nor fully opaque—Sonke Johnsen of Duke came in to lecture on underwater translucence. When another expert, Mike Graham of Moss Landing Marine Laboratories, mentioned that kelp does not grow on coral reefs, Stanton ordered all the designs for the coral-reef sequences to be redone sans kelp.

The look of the ocean itself presented a new challenge. Early in the technical development process for the film, a team under supervising technical director Oren Jacob set out to isolate the visual cues that marked a realistic undersea setting. They concluded that from an audience's perspective, those cues are lighting that shines down through the water in distinct beams, lighting that shifts about on the ocean floor, small bits of floating debris, the constant undulating of plants from the surge and swell of the water, and the fading of colors with distance on account of the water's murk. They then created a series of software tools to create the various effects they had identified, starting with a modified version of Fizt—the fur-and-cloth program from *Monsters, Inc.*—adapted to simulate the water's movement.

After Stanton saw test footage, he realized that the results looked *too* real. The technical team had duplicated the underwater world so masterfully that it was all but impossible to distinguish the computer-animated tests from the real thing. The trouble was, talking fish would seem thoroughly out of place in a photorealistic ocean. The engineers tweaked the tools to fall back to so-called "hyper-reality"—the term at Pixar for a stylized realism that had a lifelike feel without actually being photorealistic.

To make the actual production manageable, the technical directors divided into six teams, each of which dealt with the set modeling, shading, lighting, water effects, and other needs of a different part of the film. The Reef Unit handled the coral reef in which the early part of the film was set. The Sharks/Sydney Unit handled the submarine scene, the inside of the whale in which Marlin and Dory become trapped, and the above-water scenes in Sydney Harbor. The Tank Unit handled the scenes in the dentist's fish tank. The Ocean Unit handled the turtle drive in the East Australian Current, the jellyfish sequence, and the deep-sea anglerfish chase, among others. The Schooling/Flocking Unit created crowd scenes of fish and birds and collaborated on the turtle drive. The Character Unit created the nearly 120 sea creatures, humans, and bird characters.

Finding Nemo released on May 30, 2003, and went on to dislodge *The Lion King* of Disney's Katzenberg era from its throne as the highest-grossing animated film at the box office. (In inflation-adjusted dollars, however, *The Lion King* retained its pride of place, both domestically and worldwide.) It won the Academy Award for best animated feature. Kenneth Turan of the *Los Angeles Times* suggested that with "five successes out of five attempts," the company was "now the most reliable creative force in Hollywood." Chris Suellentrop of *Slate* aptly noted that with Pixar's latest hit, its record had become a phenomenon in itself: "*Finding Nemo* is Pixar's 500th home run, its 3,000th hit, its third consecutive championship."

Production of *Finding Nemo* was punctuated in September 2002 by a visit from the Japanese animation director Hayao Miyazaki, intensely popular within Japan and renowned among animation connoisseurs elsewhere for his films *The Castle of Cagliostro, My Neighbor Totoro,* and *Princess Mononoke,* among others. As a young animator, Lasseter had taken inspiration from Miyazaki's ability to entertain the whole audience, both adults and children, not just younger viewers. Lasseter had met him some two decades earlier when Miyazaki visited the Disney Feature Animation studio. Later, when Lasseter was in Japan for an animation festival in 1987, he went to Miyazaki's studio, Studio Ghibli, and showed him *Luxo Jr.* and *Red's Dream;* Miyazaki, a skeptic of computer animation at the time, was impressed with Lasseter's craftsmanship. A friendship took hold between the men when Lasseter returned in 2000 during a press tour for *Toy Story 2.*

Now, in 2002, Lasseter was serving as executive producer of Disney's English-language release of Miyazaki's *Spirited Away* (*Sen to Chihiro no Kamikakushi*), which Lasseter hoped would spark interest in Miyazaki's work among American audiences. On the morning of September 13, several days after the U.S. premiere of *Spirited Away* at the El Capitan, Lasseter brought Miyazaki to Pixar for "Miyazaki Day." The studio's cavernous atrium was hung with oversized *Spirited Away* banners. As Lasseter explained the building's design, the white-haired, gray-suited sixty-one-year-old director watched impassively as young employees zipped past on scooters.

Lasseter then took him and his interpreter around to the animators' imaginatively decorated workspaces, culminating with room 557, the office of Andrew Gordon. Gordon, an alumnus of the Looney Tunes unit of Warner Bros., had noticed upon moving in that a small door inside his office led to an empty alcove; it was meant to be an air-conditioning shaft. The space had since become something different.

As Miyazaki looked on, Gordon slid an armchair out of the way to reveal the low-slung entrance to what was now room 557.1. Miyazaki, curious, crawled through the door. Lasseter followed and showed him that the cramped space inside had been converted into a retro-style lounge—the LOVE LOUNGE, as a small neon sign proclaimed—its walls festooned with leopard- and zebra-print fabric, strings of little blue lights, tasteful pinup-girl photos, and a vermouth poster. Early rock-and-roll music came on. Finding himself in a place of pure imagination, Miyazaki finally relaxed and laughed appreciatively.

Lasseter next took the visitors to see his own office, stacked from floor to ceiling with shelves of toys. On a section of one wall was Lasseter's "Miyazaki shrine," as he called it, with a tall stuffed figure of one of Miyazaki's characters and a poster for *My Neighbor Totoro.* After lunch, Lasseter hosted a company-wide screening of Miyazaki's short film *Mei and the Kittenbus* in the studio's in-house theater. The employees watched with laughter and gasps of delight, then gave the film a standing ovation as the lights came up.

The following morning, the two men—both of them lovers of vintage cars—met at a Sonoma parking lot where they took turns driving Lasseter's open-topped 1952 Jaguar XK120 roadster. Lasseter then drove him to Sonoma Valley Airport, where the Vintage Aircraft Company was waiting to give them flying tours of the countryside in Boeing-built Stearman biplanes. Miyazaki and his producer went up in a bright red aircraft, Lasseter in a green-and-yellow one, with pilots who took them a few hundred feet up and made lazy arcs over the area's verdant vineyards.

That afternoon, Lasseter took Miyazaki and his producer and interpreter to the Lasseter family vineyard in nearby Glen Ellen. Sonoma was town; Glen Ellen was country. There, Miyazaki met Lasseter's parents, who had been living on the property for about a year. His diminutive eighty-four-year-old mother, Jewell, now retired after thirty-eight years of art teaching, startled Miyazaki with the power of her handshake. ("She's from Arkansas," her son explained. "She's strong!") His father, Paul, from whom he appeared to have inherited his hail-fellow-well-met manner, was amused to hear Lasseter describe the Japanese director grandly as one of his greatest influences.

"Really?" he said. "Glad *somebody's* been an influence on him."

The next day found Lasseter and Miyazaki back at Pixar, this time to take part in a charity benefit. The event was a *Spirited Away* screening to raise money for the Juvenile Diabetes Research Foundation, a cause that the Lasseters had taken up after one of their sons was diagnosed with the disease. At a reception after the screening, Miyazaki was approached by another Pixar director, a clean-cut man who appeared to be in his mid-twenties, but who in fact was two decades older.

"So you got to see a little bit of my story reels yesterday," Brad Bird remarked.

Lasseter confirmed that he had shown Miyazaki part of the story reels for Bird's film, which would be Pixar's next release after *Finding Nemo* came out in the spring.

Bird, in a rare moment of diffidence, followed up with a touch of anxiety in his voice. "They make any sense?" he asked. "Or did they just seem like American nonsense?"

Miyazaki replied genially, through an interpreter, "I think it's a very adventurous thing you are trying to do in an American film."

Hayao Miyazaki was prone to understatement.

Even as Pixar's reputation and commercial fortunes were climbing with the successes of its first several films, Lasseter had been plotting a radical change of course for its sixth feature. In Hollywood, where success and failure often seemed to defy logic (where "nobody knows anything," as the screenwriter William Goldman famously asserted), the ordinary response to success was to turn out more of the same for as long as possible until audiences finally became saturated with it. Yet in the spring of 2000, during production of *Monsters, Inc.* and the early stages of *Finding Nemo,* Lasseter discarded Pixar's mold twice over: first by hiring an outside director (up to then, the company's directors had all risen through the ranks), and then by embracing that director's concept of a story in which all the characters were human—or superhuman.

Brad Bird was born in Kalispell, Montana, in 1957 and raised in Corvallis, Oregon. Interested in animation from an early age, he started his first amateur effort—a retelling of the fable "The Tortoise and the Hare"—when he was eleven years old and finished it two years later. He corresponded with Walt Disney Studios. Along with like-minded students who were scattered across the country, he received an invitation while he was a student at Corvallis High School to apply to the new character animation program at CalArts.

Lasseter and Bird, classmates there, shared a belief that animation could amount to more than the children's fare it had become. "Brad would hang out all night talking about Scorsese and Coppola and how he could do what they did in animation," Lasseter remembered.

Like Lasseter, Bird went to work for Disney after college and was disappointed by what he found. Like Lasseter, he was forced out. He briefly hung out at the Lucasfilm Computer Division in the early 1980s and sought work there. Unsure where to turn after Disney, he put some of his savings into a test film to show off a few of his ideas for animated films. One of them, *Family Dog,* about a beleaguered canine in a dysfunctional household, came to the attention of Steven Spielberg, who hired Bird to write episodes of his television series *Amazing Stories.* In the 1990s, Bird contributed to the animated shows *King of the Hill, The Critic,* and *The Simpsons.*

Hollywood brought him no end of frustrated hopes. Amid his television work, he aspired to put a film of his own on theater screens, and so he did battle constantly with the nerve-wracking start-and-stop development process typical of Hollywood studios. He was briefly engaged in 1993 as the director of a planned New Line Cinema live-action comedy, *Brothers in Crime.* In the mid-1990s, he pitched several films to the animation unit of Ted Turner's media empire, one of many outlets where he tried to get a project going after his work on *The Simpsons.* He conceived a futuristic, film noir–influenced story he called *Ray Gunn* and also sought to make a film version of Will Eisner's comic-book series *The Spirit.*

"I kept having these movies get on the runway, then they would never get cleared for takeoff," he said. "My guy would get fired. Then, of course, the new guy wouldn't want to deal with something the old guy had done. Or a film that was vaguely like something I was working on would tank at the box office."

Others with whom he had been in school were outstripping him. After his personal film *Family Dog* became the basis for an episode of *Amazing Stories,* and then a spin-off series, Bird was irked to see series co-producer Tim Burton credited for *Family Dog* in media stories while his own part in it was ignored.

A 1995 deal with Turner Pictures appeared to hold the possibility of the directing breakthrough he had been waiting for. Nothing in Bird's career would be so simple, however. When the Turner companies were absorbed by Warner Bros. during a 1996 merger, it seemed he had hit still another dead end. The production executive with whom he was working at Turner left for another company.

Yet in the aftermath, his breakthrough did emerge: Warner Bros. offered him the chance to film another story that it had optioned, a

Press Release

TURNER PICTURES INKS PRODUCTION DEAL WITH WRITER/DIRECTOR BRAD BIRD

JAN. 16, 1995

Turner Pictures (TP) has signed a non-exclusive, first-look production deal with Brad Bird, it was announced Monday by TP President Dennis Miller.

Under the terms of the two-year agreement, Bird will develop and produce animated feature films for TP. . . .

Bird noted: "Turner has offered me the freedom to develop stories and styles unlike anything done before. We share the same vision of creating projects that fall outside the realm of conventional animation."

Bird has most recently served as an executive consultant and director for Fox's "The Simpsons," including the music video "Do the Bartman," which peaked as both the No. 1 single and music video in the country. He also wrote, directed and produced the original "Family Dog" for Steven Spielberg's NBC television series "Amazing Stories."

With Francis Ford Coppola and George Lucas, Bird co-wrote "Captain EO," a 70mm 3D musical film starring Michael Jackson. He also co-wrote the Spielberg-produced feature "Batteries Not Included."

children's book by the English poet Ted Hughes. Published in 1968 as *The Iron Man,* it offered an antiwar message through the tale of a giant metal-eating robot and his friendship with a farmer's son. Pete Townshend of the band The Who had made an album based on the book in 1989, and this in turn had led to Warner Bros. optioning the book. (Hughes followed up with a little-read 1993 sequel, *The Iron Woman,* an anti-pollution allegory.)

Bird took the directing assignment and combined traditional cel animation with 3-D computer animation of the robot. He ratcheted up the story's antiwar theme by setting the film in the America of the late 1950s, a time of Cold War fears, and by changing the story's villain: an otherworldly "space-bat-angel-dragon" in the book, a U.S. government agent in the film. Although Warner Bros. gave Bird half the budget of a Disney animated feature, and a year's less time, the film that he turned in—*The Iron Giant*—was widely considered a masterpiece.

Unfortunately for him, the feature-film business of Warner Bros. Animation was in flux at the time of the film's release in 1999, and the studio did little to promote it. The theatrical run of *The Iron Giant* consisted for the most part of empty and nearly empty cinemas. Bird was heartsick.

He still had the stories he had created during his period of knocking on doors in Hollywood. One of these featured a family of superheroes obliged to go into hiding by living as ordinary citizens. He imagined it as an homage to the 1960s comic books and spy films of his boyhood, to be brought to the screen in 2-D cel animation. Personal issues, however, had percolated in the story as they weighed on him in life. At the time he was drafting the story in the mid-1990s, he was in his late thirties. He and his wife had just had their third child, and he was wondering, with a measure of fear, about the conflict between career and family responsibilities. He had very high aspirations for his filmmaking—but were such aspirations attainable only at the price of your family life, a worse-than-Faustian bargain? If you hadn't succeeded in your career, really succeeded, by the time you were middle-aged with commitments at home, did you have to reconcile yourself to scaling down your definition of success, or giving it up altogether? All these ponderings seeped into the superhero story.

After it became obvious that *The Iron Giant* would not find the audience Bird was hoping for during its theatrical run, he reconnected with Lasseter and pitched his superhero idea. Lasseter was sold and the company announced a multifilm contract with Bird on May 4, 2000.

Bird came to Pixar with the lineup of the story's family members already worked out: a mom and dad, both suffering through the dad's midlife crisis; a shy teenage girl; a cocky ten-year-old boy; and a baby boy. Bird had based their powers on family archetypes.

> The dad is always expected in the family to be strong, so I made him strong. The moms are always pulled in a million different directions, so I made her stretch like taffy. Teenagers, particularly teenage girls, are insecure and defensive, so I made her turn

invisible and turn on shields. And ten-year-old boys are hyperactive energy balls. Babies are unrealized potential.

In Bird's story, the golden age of superheroes—a time when supermen and superwomen safeguarded ordinary citizens from mishap and had their accomplishments celebrated in newspapers and magazines—has come and gone. Because the government has forced all superheroes to turn away from their lives of heroism, the film's family members (except the baby) are cut off in some way from a major part of themselves. Bob Parr—formerly Mr. Incredible—is living in the past, dwelling on glory days, while disengaged from both his family and his boring insurance company job. Helen Parr—the former Elastigirl—is conversely living only in the present, believing all will be well if she treats her superhero past as if it never happened. Daughter Violet and son Dashiell ("Dash") have grown up forbidden to use their special powers. Even the Parrs' adopted surname is a proclamation of their supposed averageness.

The instruments of the family's recovery—of their connecting with their true selves—are two strikingly unorthodox characters. One of them, Edna Mode, is the former costume designer of choice for the superhero elite and must now content herself with designing for mere supermodels. ("I used to design for *gods,*" she fumes.) Bird envisioned her as having a scientific and engineering background and consequently decided she would be half German and half Japanese, nationalities that he felt were the embodiment of technical achievement.

After several failed attempts to cast Edna Mode, Bird took on her voice role himself. It was an extension of the Pixar custom of tapping in-house staff whose voices came across particularly well on scratch dialogue tracks. Story artist Bob Peterson had been the voice of the officious clerk Roz in *Monsters, Inc.* and of Mr. Ray, the singing schoolteacher, in *Finding Nemo;* Andrew Stanton had been the voice of Crush, the surfer-dude turtle, in *Nemo* and Emperor Zurg in *Toy Story 2;* Joe Ranft provided the voices of Jacques the cleaner shrimp in *Nemo,* Heimlich the caterpillar in *A Bug's Life,* and Wheezy the penguin in

Toy Story 2. On *The Incredibles* itself, in addition to Bird's role, animators Bret Parker and Bud Luckey were respectively the voices of Kari the babysitter and government functionary Rick Dicker.

The other major figure in bringing the lives of the Incredibles back on track was Buddy Pine. As a boy, he was a gifted inventor who idolized Mr. Incredible. After Mr. Incredible curtly rejected him, his admiration turned to scorn and hatred of all superheroes. As an adult with the nom de guerre of Syndrome, he would become a fearsome threat to the Parr family, and in doing so he would force Bob and Helen to embrace both their present and their past. (Bird's original pitch focused on a generic urbane villain named Xerek; Syndrome, a secondary villain, was to die early in the film. When Lasseter heard the pitch, he liked Syndrome better than Xerek, and thus Syndrome moved front and center.)

Bird was a departure from other Pixar directors in more ways than one. Not only was he an outsider, he brought an auteur approach not found in earlier Pixar productions. Where Pixar films typically had two or three directors and a battalion of screenwriters, *The Incredibles* had *one* director and *one* writer, and Bird was it—in substance as well as in name. Even his manner of using storyboards had a benignly autocratic touch. Other Pixar directors relied on storyboards simply to relate the dialogue and emotions of a scene, leaving it to other departments to work out the details: the blocking of characters' movements, lighting, and camera moves. On *The Incredibles,* as on *The Iron Giant,* Bird insisted that the storyboards themselves define those details, which the other departments would then carry out.

For the technical crew members, the human characters in *The Incredibles* posed a difficult set of challenges. While humans had had supporting roles in the two *Toy Story* films, in *Finding Nemo,* and in *Monsters, Inc.,* Bird's was the first Pixar film to be populated entirely by human beings. Moreover, Bird would tolerate no compromises for the sake of technical simplicity. Mr. Incredible *had* to be muscular. Where the technical team on *Monsters, Inc.* had convinced Pete Docter to accept pigtails on Boo to make her hair easier, Violet *had* to have long hair that obscured her face; it was integral to her character.

The skin of the characters gained a new level of realism from a technology to produce what is known as "subsurface scattering." Human skin is not fully opaque; part of what makes it look like skin is that it allows some light to reach its inner layers and scatter among them before reflecting back. Consequently, skin looks unnatural if it is rendered as an ordinary solid surface. Algorithms to recreate subsurface scattering, pioneered by a Stanford researcher named Henrik Wann Jensen, allowed the technical crew to mimic human skin more effectively.

Yet the humans' skin could not be *too* realistic. It was well known that as depictions of humans became more lifelike, audiences would perceive them as more appealing—until the realism reached a certain point, close to human but not quite, when suddenly the depictions would be perceived as repulsive. The phenomenon, known as the "uncanny valley," had been hypothesized by a Japanese robotics researcher, Masahiro Mori, as early as 1970. No one knew precisely why it happened, but the sight of nearly human forms seemed to trigger some primeval aversion in onlookers. Thus, the minute details of human skin, such as pores and hair follicles, were left out of *The Incredibles'* characters in favor of a deliberately cartoonlike appearance.

The characters' designs came from Tony Fucile and Teddy Newton, whom Bird had brought with him from Warner Bros. For the technical team members who would have to translate their drawings into computer models, and then rig the models with control points to define their movement, it was important to develop a clear understanding of the characters' anatomies. The models would loosely imitate human structure, with skin, muscle, and fat attached to bones and interacting with one another. Where the crew of *Finding Nemo* had dissected fish in the service of model making, however, the crew of *The Incredibles* would have to satisfy itself with copies of the medical school textbook *Gray's Anatomy.*

The challenges of *The Incredibles* did not stop with modeling humans. "The hardest thing about *The Incredibles* was there was no hardest thing," supervising technical director Rick Sayre said afterward. "Brad ordered a heaping helping of everything on the menu. We've got it all: fire, water, air, smoke, steam, explosions, and, by the way, humans. . . . Getting hair to work at all and to move and clothing and then doing it with a big ensemble cast. It's a Pixar compendium."

As production proceeded, Bird became accustomed to seeing "the Pixar glaze," as he called it, "where these complete technical geniuses would just grow pale and look at each other like, 'Does he know what he's asking?' "

At times, Bird's reach exceeded even Pixar's grasp. Bird decided that in a shot near the end of the film, baby Jack-Jack would undergo a series of transformations—five, to be exact—in one of which he would turn himself into a kind of goo. Sayre reckoned it would take a group of technical directors two months to work out the goo effect, and production was at a point when two months of their time was indescribably precious. Sayre petitioned the film's producer, John Walker, for help. Bird, who had brought Walker over from Warner Bros., took great exception to the idea that Jack-Jack could undergo a mere four transformations and that *The Incredibles* could do without the goo-baby. They argued over the issue in meeting after invective-laced meeting for two months before Bird finally gave in. "I told Brad, 'I'd love to give you goo,' " Walker recalled. " 'I want to give you goo. But four transformations are enough! *Please!*' "

The Incredibles opened on November 5, 2004, as Pixar's first PG-rated film (thanks to its action sequences). Audiences and critics alike rewarded Bird's perfectionism. *Variety* called it "the most ambitious and genre-expanding entry in Pixar's extraordinary run of innovative and monstrously successful computer animated pictures," and observed, "As deliriously smart escapist fare, 'The Incredibles' is practically nonpareil."

Many comic fans assumed that *The Incredibles* had been influenced substantially by Alan Moore's graphic novel *Watchmen.* In the world of *Watchmen,* as in *The Incredibles,* costumed heroes have been outlawed and someone is killing them off. Syndrome has a parallel in the *Watchmen* character Rorschach, a red-haired killer who had been

The Incredibles producer John Walker, director Brad Bird, and executive producer John Lasseter reveal their superhero identities at a party following an October 24, 2004, invitation-only screening of the film in Los Angeles. *Kevin Winter/Getty Images*

rejected in his boyhood by a beloved figure. Although Rorschach is a vigilante rather than a pure villain like Syndrome, it is tempting to infer that Bird's character is based on Moore's. In both *The Incredibles* and *Watchmen,* moreover, a character rails against the stupidity of capes. There was another seeming influence on *The Incredibles,* as well: The plight of the boy Dash, forbidden from excelling through the use of his extraordinary powers, appeared at first glance to derive from Kurt Vonnegut's story "Harrison Bergeron." In interviews, Bird explained that the resemblances were coincidental and that he had read neither work.

The film won the Academy Award for best animated film, up against two DreamWorks nominees (*Shrek 2* and *Shark Tale*); it was Pixar's second straight victory in the category. Film aficionados admired Bird's having taken the well-established conventions of superhero storytelling and made them fresh.

Of note were the reactions of the widely read film Web site AintItCool.com, run by the then-thirty-three-year-old film fanatic Harry Knowles, whose gung ho enthusiasm resembled that of the young Buddy Pine, and whose flowing red hair lent him a family resemblance to Syndrome. ("Brad Bird is a SUPER GENIUS," Knowles wrote when the film's trailer appeared on the Internet in mid-September. "We should all worship him. There should be monuments . . . recarve Rushmore into a giant likeness of his beautiful magnificent head!") One of the site's resident critics, "Moriarty" (Drew McWeeney), rhapsodized, "I've seen the film multiple times now, and there are moments that are burned into my memory, those indelible goosebump moments that make us fall head-over-heels in love with a movie." He called Bird's script "as sharp and incisive as the best work of Billy Wilder and I. A. L. Diamond," and cited Bird's "impeccable ear for dialogue and the way a family really behaves." Another, "Quint" (Eric Vespe), told fans the film was "everything you hoped for. . . . This film is going to make so much money, I don't know if there'll be enough bean counters in Hollywood to take on the challenge of counting it all."

Knowles himself was uncharacteristically silent on *The Incredibles* for almost two weeks before weighing in. "I've not written about the film, simply because—oddly—I can't think of what to say."

> It's perfect. It's just an absolutely perfect film. . . . Yeah, I've been getting e-mails out the wazoo claiming that SYNDROME is me—and I'm totally OK with that. I haven't a clue if it's true, but when Syn exclaims that he's "Geeking Out," I have to admit—I was "Geeking Out" too. . . . Brad Bird has now made two of the best animated . . . strike that, just two of the best films period of the decade.

While Pixar enjoyed a sixth triumph in *The Incredibles,* Jobs was embroiled in a public feud with the head of its distribution partner, the Walt Disney Co. The outcome of their feud would recast Pixar's future and the future of Disney animation.

11. HOMECOMING

Cars, Ratatouille

By the late 1990s, with the box-office numbers of Pixar and DreamWorks Animation in view, Walt Disney Co. chief executive Michael Eisner had come to believe that the future of animation was in 3-D computer graphics. Accordingly, Disney built a state-of-the-art computer animation and special-effects studio, named the Secret Lab, in a converted Lockheed aircraft plant near Burbank Airport. The first project of the new studio, *Dinosaur,* inserted talking computer-generated dinosaurs into real landscapes; it did respectably well when it opened in 2000, but its performance failed to justify its exorbitant production costs. It was a poor omen for Disney's first attempt at independence from Pixar in computer animation.

The second project of the Secret Lab was called *Wild Life;* it was to be a brash film, by Disney standards, about youths who spent their nights partying and dancing at clubs. The studio spent millions of dollars on preproduction, commissioning script after script, extensive storyboarding, and elaborate test footage.

Old Disney hand Floyd Norman joined the *Wild Life* story team and was puzzled by the choice of subject matter. The tests were technically impressive, but the scenario seemed far afield from Disney family fare. The script was laden with coarse sexual innuendo. Did the producers and directors even realize who employed them? he wondered. It wasn't DreamWorks, it wasn't Warner Bros., it wasn't Columbia—it was the Walt Disney Company.

In August 2000, Roy Disney, who was now in his sixteenth year as chairman of Disney Feature Animation, flew in from his castle in Ireland for a routine viewing of story reels for the film. Appalled, he promptly shut the project down. The following year, the company closed the Secret Lab.

Disney was faring little better with its 2-D cel-animated features. A string of releases—*The Emperor's New Groove* (2000), *Atlantis: The Lost Empire* (2001), and *Treasure Planet* (2002)—brought disappointment at the box office. Only Disney's production agreement with Pixar, along with a few live-action hits like *Pearl Harbor,* was buoying Walt Disney Studios.

While Disney and Pixar continued to prosper from their relationship, tensions inevitably arose between their chief executives. The men's backgrounds could hardly have been more of a contrast—Eisner, brought up with every advantage as the son of an old-money Park Avenue family; Jobs, the adopted son of lower-middle-class parents; Eisner, the career executive; Jobs, the ex-hippie. Yet the true root of their conflicts was neither their differing backgrounds nor the bread-and-butter disagreements involved in doing business together. It was in their similarities: Besides being notably aggressive in representing their companies' interests, each man was stubborn to the point of petulance and prone to taking disagreements personally.

Throughout the 1990s, the two had successfully managed the conflicts in the business needs of the two companies, at least on paper. Eisner had acceded, in 1997, to Jobs's demands for the new contract that gave Pixar co-ownership of its films and a higher take of the profits. Later, Jobs had unhappily given in to Eisner's insistence on sticking to the letter of that contract, which excluded *Toy*

Story 2, as a sequel, from counting toward the five Pixar films that Disney had been assured.

A turning point in their relationship came after the release of *Toy Story 2,* when Eisner, pleased with the sequel's revenues, wanted Pixar to make a *Toy Story 3.* Jobs argued that Disney had already gotten one bonus film, so to speak, in *Toy Story 2.* He observed that one of the rationales for treating sequels differently—that they fared worse at the box office—no longer held true: *Toy Story 2* had done far better than *Toy Story.* Eisner, for his part, had the terms of the contract in his favor, and wanted as many fully original Pixar films as possible to generate new character franchises for merchandise and theme-park attractions. Jobs was intensely frustrated by Eisner's immovable position that if Pixar made a third *Toy Story* film, it would be treated like *Toy Story 2,* effectively stretching the five-film deal of 1997 into a seven-film deal. Not only had Eisner riled Jobs, he had antagonized Lasseter, who wanted to make another *Toy Story* film.

Relations between Eisner and Jobs took still another turn for the worse when, on the morning of February 28, 2002, Eisner testified before the Senate Committee on Commerce, Science and Transportation regarding digital piracy of films. Although Eisner's prepared statement to the committee was mostly couched in bland bureaucratic prose, he loosened up in the question-and-answer session afterward. "The 'killer app' for the computer industry is piracy," Eisner asserted. "They think their short-term growth is predicated on pirated content."

One technology company, in Eisner's estimation, was especially blatant in promoting piracy:

> There are companies, computer companies, that their ads—full-page ads, billboards up and down San Francisco and L.A., that say—what do they say?—"Rip, mix, burn." . . . They are selling the computer with the encouragement of the advertising that they can rip, mix, and burn. In other words, they can create a

theft and distribute it to all their friends if they buy this particular computer.

It was unmistakably an attack on Apple and its "Rip. Mix. Burn." advertising campaign. Eisner had evidently confused the term *rip*—meaning, simply, to copy music or video from a compact disc or DVD into a computer, whether for lawful or unlawful purposes—with *rip off.* Thomas Schumacher, president of Disney Feature Animation, soon got a phone call from an agitated Steve Jobs, who had taken the comments as a personal accusation. "Do you know what Michael just did to me?" Jobs demanded.

Jobs continued to bristle over *Toy Story 3,* Eisner's testimony, and what Jobs saw as Eisner's high-handed manner of dealing with Pixar. That summer, using studio head Dick Cook as a go-between, he reached out to Roy Disney and vented his frustrations over a private dinner.

Roy was a receptive audience. He had led the boardroom coup that put Eisner in power back in 1984, and had admired Eisner's accomplishments in revitalizing the company during the decade when he had Frank Wells and Jeffrey Katzenberg as his subordinates. Since then, however, his esteem for Eisner had been waning. Roy was troubled by what he regarded as Eisner's proclivity for micromanagement, his nickel-and-diming of the Disney theme parks (especially the new Disney's California Adventure), and his general overemphasis on short-term profit. Beyond that, he had come to distrust Eisner personally since learning that Eisner had placed a mole in Disney Feature Animation to keep tabs on Roy's actions.

Jobs spoke freely to Roy about his grievances. His punch line was startling: After Pixar fulfilled its obligations under the 1997 deal, there would be no more Pixar films for Disney—not while Eisner was in charge. "I'll never make a deal as long as Eisner is there," Jobs told him.

Roy and his business partner Stanley Gold were on the Walt Disney Co. board. (Gold ran Roy's personal investment firm, Shamrock Holdings.) They told their fellow directors that they thought the

Pixar relationship was at risk—the outcome that Jobs surely had calculated when he sought Roy out. Not wishing to incur Eisner's wrath, they held back the fact that Roy had actually met with Jobs.

Eisner gave the impression, at least, of being unconcerned. In an e-mail to board members on August 22, he related his impressions of the company's upcoming films. On *Finding Nemo,* due for release the following summer, he was dismissive, as if rooting for Pixar to get its comeuppance. "Yesterday we saw for the second time the new Pixar movie 'Finding Nemo' that comes out next May. This will be a reality check for those guys. It's okay, but nowhere near as good as their previous films. Of course they think it is great."

As it turned out, the film would become the highest-grossing animated movie in history and would win the Academy Award for best animated feature.

In the spring of 2003, Jobs offered Eisner terms for a new deal—terms so one-sided that they seemed less like an opening offer than a prelude to an outright breakup. Under Jobs's proposal, Disney would no longer have co-ownership of Pixar's future films, as it did of the films made under the 1997 agreement. Disney would no longer receive a split of the profits from Pixar's future films, but only a distribution fee of 7.5 percent. Disney's exclusive authority to distribute the films would last only five years.

Jobs insisted, moreover, that the final two films of the 1997 deal, films not yet made—*The Incredibles* and *Cars*—be transferred to the new deal. Under his proposal, Disney would receive a distribution fee of 10 to 15 percent on those films, but not a dollar of the profits otherwise. With that point alone, Disney would be giving up rights it already had in those films, worth hundreds of millions of dollars, in return for the privilege of distributing Pixar's subsequent works. Jobs's one concession to Disney was to allow the free use of Pixar's future characters in Disney's theme parks, as Disney was doing with Pixar's existing characters.

Pixar's four consecutive hit films at that point had cumulatively grossed more than $1.7 billion at the box office worldwide and had sold more than a hundred million DVDs and tapes. Audiences trusted the Pixar name. (Jobs's insistence on strong Pixar branding in the 1997 deal had paid off.) With its sterling record, no debt, and more than $500 million in cash, Pixar was less dependent on Disney than ever. Any of a number of major studios, as Jobs saw it, could provide strong marketing and distribution. By May, Jobs had met with executives of Sony Pictures Entertainment and the parent companies of Warner Bros. and Twentieth Century Fox. (Fox was perhaps an unlikely choice given that it also owned Blue Sky Studios, a Pixar rival that had made the computer-animated *Ice Age.*)

Eisner still had a card to play, however. Under the 1991 and 1997 agreements, Disney owned *Toy Story* and its characters entirely, and also had the right to make sequels to any of Pixar's other films—with or without Pixar's involvement. The idea of Disney cranking out *Toy Story 3, Finding Nemo 2,* and the like drove Lasseter to distraction. He regarded the films almost as his children, and there was little reason to expect that Eisner would tend them with any sort of care. Disney-made sequels under Eisner, it seemed, would be objects of commerce above all.

"These were the people that put out *Cinderella II,*" Lasseter later said mordantly, referring to the 2002 direct-to-video sequel.

As he had with Roy Disney, Jobs continued to work through back channels into Disney's board. During the summer of 2003, after former vice president Al Gore joined the board of Apple Computer, Gore contacted Disney board member George Mitchell, the former Senate majority leader, to relay Jobs's frustration with Eisner. Mitchell queried Eisner about the progress of the Pixar talks. Eisner was irked when he learned afterward about Gore's involvement.

The episode was no more than an annoyance; Eisner had the board's confidence and its loyalty. Roy Disney and Stanley Gold were its only dissidents, regularly venting their criticisms of Eisner's compensation and performance to the other directors. Roy would soon be on his way out the door, however, one way or another. The news would be delivered by John Bryson, a utility executive who had joined the board several years earlier.

A week before Thanksgiving, over drinks at a wine bar, Bryson stunned Roy by informing him that when his current term ran out, he would not be permitted to serve again. Roy had passed the board's mandatory retirement age of seventy-two.

But that rule was for *outside* directors, Roy reminded him. It didn't cover executives who were on the board. He was still chairman of Disney Feature Animation.

The committee had already met and decided, Bryson told him—referring to the board's governance committee, of which Bryson was chairman.

Roy was aghast. After conferring with his wife and children, and with Gold and other trusted counselors at Shamrock, he tendered his resignation from the board and from his job on November 30, 2003. He arranged to have his resignation letter hand-delivered to Eisner at the same time it was being faxed to the national media.

"You had a very successful first 10-plus years at the company in partnership with Frank Wells, for which I salute you," Roy wrote. "But, since Frank's untimely death in 1994, the Company has lost its focus, its creative energy, and its heritage."

Eisner's dealings with Pixar were part of a larger set of failures that Roy cited.

> As I have said, and as Stanley Gold has documented in letters to you and other members of the Board, this Company, under your leadership, has failed during the last seven years in many ways:
>
> 1. The failure to bring back ABC Prime Time from the ratings abyss it has been in for years and your inability to program successfully the ABC Family Channel. . . .
>
> 2. Your consistent micro-management of everyone around you with the resulting loss of morale throughout the Company.
>
> 3. The timidity of your investments in our theme park business. At Disney's California Adventure, Paris and now in Hong Kong, you have tried to build parks "on the cheap" and they show it and the attendance figures reflect it.
>
> 4. The perception by all of our stakeholders—consumers, investors, employees, distributors and suppliers—that the com-

pany is rapacious, soul-less, and always looking for the "quick buck" rather than long-term value which is leading to a loss of public trust.

5. The creative brain drain of the last several years, which is real and continuing, and damages our Company with the loss of every talented employee.

6. Your failure to establish and build constructive relationships with creative partners, especially Pixar, Miramax, and the cable companies distributing our products.

7. Your consistent refusal to establish a clear succession plan.

Roy concluded, "Michael, it is my sincere belief that it is you who should be leaving and not me."

An open letter from several Disney Feature Animation employees—an animator, a screenwriter, and a director—quickly appeared online in support of Roy. It drew thousands of other signers from the ranks of current and former Disney animation employees in the months that followed. Under Eisner, the writers complained, "The unique traditions of visual storytelling, humor and personality animation on which the Walt Disney Studio had thrived, gave way to politically correct sloganeering, stale one-liners and film seminar formulae to which audiences have failed to respond."

Gold submitted his resignation several days after Roy's. Now outsiders to the company that Roy's father had co-founded, the two men pondered how to defenestrate the executive they had helped to install nineteen years earlier.

Disney's and Pixar's differences regarding their next distribution deal, meanwhile, remained unresolved. On the afternoon of Thursday, January 29, 2004, about ten months after giving his contract demands to Eisner, Jobs made a surprise announcement that he was ending negotiations. On Disney's behalf, Dick Cook gamely put the best face on the situation. "No one has a lock on talent," he told *The Wall Street Journal.* "No one has a lock on creativity or technology or storytelling."

While that was true, it was also a fact that Disney's in-house feature films had numbered only one significant success in recent years—2002's *Lilo & Stitch*—amid a series of box-office letdowns. Based on Eisner's belief that cel animation was obsolete, Disney Feature Animation had fired most of its artists and shut down most of its own animation operations—including the Orlando, Florida, studio that had created *Lilo & Stitch.* A rump operation devoted to computer animation was working on *Chicken Little* and *A Day with Wilbur Robinson* (later retitled *Meet the Robinsons*).

Eisner's own public response was genial. "We have had a fantastic partnership with Pixar and wish Steve Jobs and the wonderfully creative team there, led by John Lasseter, much success in the future," he said in a written statement the same afternoon. "Although we would have enjoyed continuing our successful collaboration under mutually acceptable terms, Pixar understandably has chosen to go its own way to grow as an independent company."

The statement also cited Disney's distribution relationships with other computer animation studios (namely, Vanguard Films, which was working on the World War II messenger-pigeon film *Valiant,* and Complete Pandemonium and C.O.R.E. Feature Animation, producing *The Wild*), its in-house computer-animated films in production, and its forthcoming release of the 2-D *Home on the Range.*

The day of the announcement, after Jobs made a courtesy call to Dick Cook to give him the news that Pixar would be walking away, he made a second call, this one to Roy Disney. Jobs reiterated to Roy that he would never make another deal with the Walt Disney Co. as long as Eisner was there. The "negotiations" and their breakoff had, in short, been pure theater, meant only to embarrass Disney's chief executive.

In Roy's mind, Eisner had come to be symbolized by the Wicked Witch of the East in *The Wizard of Oz.* "When the wicked witch is dead, we'll be together again," Roy told Jobs.

Pixar's quarterly earnings conference call was scheduled for the following week. The calls gave investment analysts from firms such as

Morgan Stanley and Deutsche Bank the opportunity to hear briefings from Jobs and his chief financial officer, Ann Mather, on the state of the company. Ever since 2000, when the SEC had laid down rules drastically cutting back on one-to-one conversations between company officers and analysts about company news, such calls had become a critical channel for chief executives to get their messages out to investors and respond to their concerns.

After Jobs and Mather went through the details of Pixar's enviable financial condition, Jobs turned to the subject of the Disney relationship. He began by quoting a *Los Angeles Times* story about Eisner's criticism of *Finding Nemo* (which had since opened and broken box-office records). Jobs read a passage of the article reporting on Eisner's

Pixar CEO Steve Jobs, Walt Disney Studios chairman Dick Cook, and Pixar senior vice president–creative John Lasseter (L–R) arriving at the premiere of *Finding Nemo* at Disney's El Capitan Theatre on May 18, 2003. Disney CEO Michael Eisner, Cook's boss, predicted that the film would disappoint and would bring Pixar its comeuppance; *Finding Nemo* instead went on to become the highest-grossing animated film in history.
Dan Steinberg/Getty Images

belief that Pixar was heading for a fall. Jobs said that he and others at Pixar had heard the same story from people at Disney.

"As you know, things turned out a little different," Jobs added sardonically.

Jobs told the group he believed there were at least four major studios besides Disney that would do an excellent job marketing and distributing Pixar's films. "I have personally received calls from the heads of each of these studios during the past five days, all expressing a very, very strong interest in working with Pixar," he said.

He added that he saw little value in Disney's creative contributions. "The truth is that there has been little creative collaboration with Disney for years," he said. "You can compare the creative quality of Pixar's last three films, for example, with the creative quality of Disney's last three animated films, and gauge each company's creative abilities for yourself."

He continued to bad-mouth Disney's recent releases when he turned to the subject of marketing.

"Marketing is important, and we have enjoyed working with Dick Cook and his talented team at Disney," he said. "But no amount of marketing will turn a dud into a hit. Not even Disney's marketing and brand could turn Disney's last two animated films, *Treasure Planet* and *Brother Bear,* into successes."

His complimentary reference to Cook was sincerely meant, from all appearances, but it was also astute politics: Cook would be responsible for the marketing of the two films that Pixar still owed Disney.

Jobs did concede one regret, namely, the prospect of Disney-made sequels to Pixar's films. "We feel sick about Disney doing sequels, because if you look at the quality of their sequels, like *The Lion King 1½* and their *Peter Pan* sequels and stuff, it's pretty embarrassing."

When Wall Street had cast its vote on Jobs's move at the time of his announcement, Pixar investors were evidently optimistic; its shares rose modestly the next day, around 3 percent, while Disney's fell about 2 percent. Still, some of the analysts struggled to understand why Jobs thought Pixar came out ahead by forsaking its highly successful relationship with the world's top marketer of family entertainment.

"It would seem to me that Disney was in the position where they almost had to say no," offered Kathy Styponias, a longtime analyst of entertainment companies at Prudential Equity Group.

> Would you not have been happier—or would you have been willing to take a higher amount on the split instead of 50–50 . . . as opposed to just a straight distribution deal? Because I'm having a hard time figuring out how the numbers . . . It's hard to see how the numbers balance out.

One might as well have looked for rationality in a playground fight. The truth was that Jobs's decision was always less about "the numbers" than about his dislike for Michael Eisner.

Eisner, meanwhile, was suddenly waging a defensive war on two fronts. One was the sixty-six-billion-dollar hostile takeover offer announced on February 11 by the cable television firm Comcast, which saw opportunity in Disney's half-dozen years of weak stock performance and in the shareholder dissension bred by Roy's highly public departure.

The other, seemingly far less threatening, was a campaign launched by Roy himself with Stanley Gold, which they called "Save Disney." Their goal was to bring about the firing of Michael Eisner—even though the board was now bereft of anyone who was even open to argument on the subject. It was too late for Roy to put forward another slate of directors at the next shareholders' meeting in March. All he could do was exhort shareholders to make a protest statement by withholding their votes for Eisner when he ran for reelection to the chairmanship of the board, along with several other board members whom Roy found particularly objectionable.

The landscape was far different than it had been when Roy overthrew Disney chief executive Ron Miller two decades before. The Bass family, major Disney shareholders who had supported Roy's efforts in 1984, had since sold their stock. (Indeed, Eisner himself was now one of the company's top individual shareholders.) Eisner had the incumbent's advantage; the resources of the company were his to deploy in opposition. Save Disney, in contrast, consisted essentially of a Web site, SaveDisney.com, and the voices of Roy Disney and Stanley Gold. Large public companies routinely swatted down shareholder activism of the scale of Save Disney.

Roy's team intuited, however, what Eisner's did not: namely, that the combination of his Disney name plus the reach of the Internet could empower Save Disney to get its story out to the grass roots of the company's customers and small investors. Shamrock Holdings had registered the SaveDisney.com domain name the day after Roy's resignation. Visitors to the Web site saw Roy's impassioned open letters railing against what he portrayed as Eisner's squandering of the Walt Disney Co. legacy. As the national news media ran articles on Roy's campaign, the site added links to those articles, creating the appearance of momentum for Roy's side and inevitable doom for Eisner's. The company's Web site, meanwhile, ignored the campaign, even on its investor relations pages, as if the company were unwilling to dignify the criticisms with a response—thus assuring that anyone relying on the Internet for their information saw only Roy's perspective on the issues.

At the same time, Roy and Stanley Gold also pursued the more traditional tactics of a proxy fight, reaching out personally to institutional shareholders and shareholder advisory firms. Their first breakthrough came when, on February 11, the influential advisory firm Institutional Shareholder Services recommended that its clients withhold their votes for Eisner, citing "the Pixar–Disney divorce" among other factors. In the weeks to follow, Roy and Stanley Gold also persuaded the mutual fund giants Fidelity and T. Rowe Price to withhold their votes, as well as the employee pension funds of California and a half-dozen other states.

On the morning of March 3, thousands of Disney shareholders streamed into the Philadelphia Convention Center, the venue of that year's shareholders' meeting. After hearing speeches from Eisner, Roy, and Gold, followed by presentations from Disney's chief financial officer and its division heads on the vibrancy of the company, the attendees received astonishing news: Some 43 percent of votes had been withheld from Eisner. It was the highest no-confidence vote ever against a chief executive officer of a major company. (A later count showed that the withhold vote was actually a little over 45 percent.)

The same evening, a chastened board announced that it had decided to separate the positions of CEO and board chairman. For the role of chairman, it had chosen George Mitchell to take Eisner's place. Yet the board was unwilling to go any farther; Eisner would stay on as chief executive, the announcement went on, with the board's "unanimous" support.

The threat of a hostile takeover soon dissipated as Comcast dropped its bid the following month. Save Disney, however, did not go away as easily. Roy and Gold continued to press their advantage and vowed to return to the next shareholders' meeting for another try. Six months after the withheld vote and the loss of his chairmanship, Eisner revealed in September 2004 that he would retire at the end of his contract term two years hence. The following March, under pressure from the board, Eisner assented to leaving a year early. On the matter of Eisner's successor, the board interviewed eBay chief executive Meg Whitman and Eisner's number two, Disney president and chief operating officer Bob Iger. George Mitchell announced on March 13, 2005, that Iger would be the next CEO of the Walt Disney Co. effective October 1.

Bob Iger, born on February 10, 1951, had dreamed in high school of becoming a news anchor like Walter Cronkite. He majored in broadcasting at Ithaca College, where he studied television writing under *Twilight Zone* creator Rod Serling. After graduating magna cum laude in 1973, he spent a year as a weatherman and reporter for a local television station. During that time, he became dissatisfied with working in front of the camera and began to doubt his abilities as a newscaster. He took an entry-level studio production job with ABC in 1974. Two years later, he joined ABC Sports and moved up through the ranks of management.

His years at ABC gave him a sink-or-swim introduction to the business side of news and entertainment. He swam. By the time Disney bought ABC's parent company, Capital Cities/ABC, in 1996, Iger was Capital Cities' president and chief operating officer. Eisner kept him on and also put him in charge of Disney's overseas operations before elevating him to the number two position in early 2000.

At fifty-four, Iger still looked the part of the trusted anchorman that he had once aspired to be. He had proven his diplomatic skills and political astuteness time and again, most recently in leading Disney's deal making with the Chinese government for the development of Hong Kong Disneyland, a strategic breakthrough for the company in the Chinese market. Disney executives regarded China and India, with their burgeoning middle classes, as its top-priority markets. A slide at the 2005 shareholders' meeting showed a composite photo of the Taj Mahal with Mickey Mouse ears in place of its dome—an expression, only mildly exaggerated, of the company's ambitions.

Iger had succeeded with the People's Republic of China, but it was unclear whether he could do the same with Steve Jobs—compared to the Chinese leaders, Jobs was far more dogmatic. Nonetheless, Iger had reason to believe he could work out a distribution deal with Pixar where Eisner had failed; Jobs had repeatedly indicated that Eisner, not the Walt Disney Co. as a whole, was the source of his displeasure. The day Mitchell announced his promotion, Iger phoned Jobs and Lasseter—not to make an offer (Eisner still had six months before retirement), but simply to check in and pay his respects.

Eisner was moving ahead with Pixar sequels. Dick Cook had disclosed at the March 2004 shareholders' meeting that Woody and Buzz "will live forever" in Disney-made sequels, starting with *Toy Story 3,* which the studio had in development. Eisner himself revealed at an industry conference a few months later that Disney also had sequels to *Monsters, Inc.* and *Finding Nemo* in development. He found it "a little bit embarrassing," he said, that DreamWorks Animation had a sequel to *Shrek* out that summer, while "for whatever reason, which I am not going to get into, we don't have the same announcements."

Disney had since assembled another digital animation facility, a successor to the now-defunct Secret Lab, specifically to create sequels to Pixar films. The facility, a former warehouse, was called Circle 7 for the Glendale street on which it was located. In March 2005, forty or so people at Circle 7 were working on *Toy Story 3,* with a script that followed Buzz Lightyear after he was recalled to the toy factory in Taiwan where he had come from. Some speculated that Circle 7 was just a ruse, a Potemkin village meant to intimidate Jobs and bring him back to the bargaining table. That theory, denied by Disney executives, became less plausible as Disney quadrupled the facility's staff over the following year.

Almost as soon as he moved into Eisner's office, Iger signaled a break with Eisner's approach to Jobs. The opportunity came thanks to Jobs's captaining of Apple Computer. Jobs's second tenure at Apple was a remarkable turnaround story, which had become famous with good reason: The year before Jobs returned as interim CEO in 1997, Apple lost more than a billion dollars. "I'd shut it down and give the money back to the shareholders," Dell Computer founder Michael Dell joked at the time. Jobs, however, had since doubled the company's sales and returned it to consistent profitability, all while maintaining its image as the epitome of cool. (Jobs dropped the *interim* from his title a little more than two years after his return.)

Impressively, some 40 percent of Apple's sales in 2005 came from a product that had not even existed four years earlier, the iPod music player. Introduced on October 23, 2001, the iPod combined mostly off-the-shelf parts on the inside with a highly refined user interface. It was not the first digital music player, but it was arguably the most intuitive and certainly the sleekest. Millions of iPod customers adored their machines in the way that earlier generations of youths fawned over their first cars. Jobs also convinced major music labels to allow Apple to sell their songs online, and so the iTunes Music Store opened on April 28, 2003, with more than 200,000 songs to start.

Iger's role—and, indirectly, Pixar's—came in the fall of 2005, when, after years of ridiculing the idea of a music player with a video screen, Jobs was ready to launch the first video iPods. The question was, where would the video content for the devices come from? Music companies had been wary of download song sales for fear of piracy and lost CD sales. Hollywood executives were, if anything, more cautious about selling digital versions of their films and television shows online. Moreover, Jobs felt constrained not to approach studios helter-skelter in pursuit of a deal; he was intent on avoiding leaks of the kind that had bedeviled him in advance of other Apple product announcements—leaks from insiders that regularly appeared on Web sites run by Apple fans. He was so perturbed by the leaks that Apple had actually sued the proprietor of one of the sites, a nineteen-year-old Harvard undergraduate who had founded the site ThinkSecret.com at the age of thirteen, and he had tried to subpoena the editor-owners of two other sites.

This time, Jobs held on to the element of surprise. After introducing the video iPods to a packed house at San Jose's California Theatre on October 12, he appeared to be winding down the presentation and then added, as an apparent afterthought, "We do have one more thing today. . . ."

It was classic Jobs showmanship. Iger joined him in front as a surprise guest; Jobs announced that the iTunes online store would offer episodes of five television shows from ABC and the Disney Channel, including the hits *Lost* and *Desperate Housewives.* It would be the first time that network television shows had been made available for purchase the day after their airing, and without commercials.

"It's great to be here to announce an extension of our relation with Apple," Iger told the crowd, then added jovially, "Not Pixar, but with Apple."

Industry analysts expressed amazement afterward that Jobs had been able to secure marquee shows for the new device. He and Iger had concluded the deal in less than a week. From Iger's perspective, clearly, it was a bank shot like Hong Kong Disneyland. The theme park in Hong Kong, possibly to be joined by a second in Shanghai, would spur demand for Disney's characters in China and, ideally, bring the government to approve a Disney-owned television channel in the world's most populous market. The license of Disney shows to Apple would deepen the relationship with Jobs and, if all went well, ease the way to a renewal of the Pixar contract. But it was still just a hope.

Lasseter had grown up in the car culture of Southern California in the 1960s and early 1970s. Closer to home, his father, Paul, was a parts manager at a Chevrolet dealership, and Lasseter took a part-time job there as a stock boy when he was sixteen. He emerged into adulthood with a typical case of Los Angeleno motorphilia; even in the early 1990s, while his means were still modest, he generally kept one car, a utilitarian one, for commuting, and another, a funky secondhand sports car, for fun. After Pixar's stock offering made him wealthy, he began building a collection that included, among other exotica, a retired NASCAR race car. Much as *Tin Toy* and *Toy Story* had been inspired by his lifelong love of toys, it was perhaps inevitable that Lasseter would eventually make a film with car characters.

Lasseter began talking with Joe Ranft, his head of story, about the idea in 1999, after their all-engulfing work on *Toy Story 2* was over. The film came a step closer to fruition with a cross-country drive that Lasseter took with Nancy and their five sons the following summer. It was two months without work and without a plan. He returned believing he had found the thread of his car film: The protagonist would learn the value of friendship and slowing down to enjoy life from day to day.

He visited raceways around the country and took racing lessons. With Joe Ranft and Jorgen Klubien, he worked on a treatment for a story set in a world without humans, where anthropomorphic cars hold jobs, have friendships and romances, and enjoy watching stock-car races. The story had a race car, an arrogant rookie, becoming separated from the tractor-trailer that was hauling him to a crucial race in California. The racer, whom Lasseter ultimately named Lightning

McQueen, was to become turned around and end up in Radiator Springs, a once-thriving town that the Interstate had bypassed and time had forgotten. The film was to be known simply as *Cars.*

In 2001, to help them create the characters of Radiator Springs and the town's appearance, Lasseter, Ranft, producer Darla Anderson, and a half-dozen other members of the production team flew to Oklahoma City, the starting point of a nine-day research trip along Route 66. Among those in the entourage were the film's two production designers, William Cone and Bob Pauley. The group drove in four white Cadillacs, led by Route 66 historian Michael Wallis, who narrated by walkie-talkie. Along the way, the group snapped photos, made sketches, talked with local residents, and even took soil samples so they could later match the color of the dirt.

"Typically, we'd go into a town and we'd hear all these wonderful stories from the locals," recalled Pauley. "We'd soak it all in while getting a haircut at the barbershop, or enjoying a sno-cone, or taking the challenge to eat a 72-ounce steak at the Big Texan."

The Big Texan Steak Ranch of Amarillo offered a free four-and-a-half-pound steak to any patron who could eat it in an hour or less. Outside the town was the Cadillac Ranch, an outdoors art installation of half-buried Cadillacs with their tailfins in the air; it would inspire a similar-looking mountain range in the film. Much of the film's natural scenery, in fact, would come from the automotive world in one way or another—from the tail-fin-shaped flowers to the overhead vapor trails in the form of tire treads.

Ranft found the basis for one character on the side of the road in Kansas, where he noticed an abandoned, rusted tow truck. In it, he saw what would become Mater, an amiable redneck who befriends Lightning and, with the town's other residents, imparts Lasseter's life lessons. (The character's name would come from that of a race fan, Douglas "Mater" Keever, a construction foreman by trade, whom Lasseter got to know during a visit to the camping area in the infield of Lowe's Motor Speedway in Charlotte, North Carolina.)

To mark his wife's birthday in September 2002, Lasseter combined research with pleasure in a nighttime drag race against her at

Sonoma's Infineon Raceway—Nancy in a motor home towing a 1960s sedan, he in an electric-yellow, Cheerios-branded NASCAR race car he had bought from Roush Racing. To much cheering from the stands, she crossed the finish line first, albeit with a generous head start.

Lasseter's belief in unrelenting research extended to the look of the film's characters. He took in the Concours d'Elegance classic car show in Pebble Beach, California, the massive North American International Auto Show in Detroit, and the design centers at Ford and General Motors. The design team spent hours studying at Fantasy Junction, an exotic-car showroom down the street from the studio. At Manuel's Body Shop, also nearby, they learned about the layering of automobile paint and inspected the surfaces of cars that had seen better days.

"While we were at Manuel's one day, we found this old beat-up chrome bumper and we asked if we could have it," remembered shading art director Tia Krater. "He started to clean it up and we said, 'No! No! Don't clean it!' It was exactly what we were looking for . . . pitting, scratches, milky blurriness, rust, and blistering. All in one bumper!"

Lightning would be voiced by the actor Owen Wilson, whom Lasseter's sons had admired since *Shanghai Noon.* Lightning McQueen was entirely a Pixar design, as were Mater and the film's villain, Chick Hicks. Most of the other vehicles were classics. Doc Hudson, the town doctor and judge, voiced by Paul Newman, was a 1951 Hudson Hornet; Ramone, the body-paint stylist and low rider, played by Cheech Marin, was a 1959 Impala; Luigi, the tire monger, played by Tony Shalhoub, was a 1959 Fiat 500; the odd couple of Fillmore the hippie and Sarge—played by George Carlin and Paul Dooley, respectively—were a 1960 Volkswagen Microbus and a 1942 Willys Jeep. Mack, Lightning's transportation, played by John Ratzenberger in his seventh Pixar feature, was a 1985 Mack Superliner.

One character in Radiator Springs was a contemporary model: the film's leading lady, Sally Carrera, a 2002 Porsche 911 voiced by Bonnie Hunt. For Sally, the studio had help from Studio Services, the Los Angeles–based film office of Porsche Cars North America. Porsche loaned 911s to the project on several occasions, once for the modeling department, once for the animators, and finally for Lucasfilm's Skywalker Sound to record the hum of its engine for the sound track. Porsche also gave Pixar a 911 for a car-customizing shop to turn into a full-scale replica of Sally, which would go on tour with a Lightning McQueen replica. Howard Buck, the head of Studio Services, suggested "Carrera" as Sally's last name.

Even the characters based on actual cars would not be literal copies—the cars of the film required eyes, a mouth, and a gender. The mouths were set into the cars' bumpers or grilles. Rather than place the eyes in their headlights, Lasseter decided that the eyes, the windows of the soul, would be in their windshields. He believed it would allow the cars to be more expressive and give them a more human appearance, while putting the eyes in the headlights would make them look vaguely snakelike. The female cars were set apart with rounder bodies and more feminine mouths (except for Lizzie, a gaunt old Model T).

Underneath the cars' skin, the modeling department built all the cars on top of the same basic rig—the same chassis and suspension of control points—adjusted for each character's size. The rig was programmed to match the cars' motion to the terrain as they moved, so the wheels would stay on the ground automatically without taxing the animators' attentions. It had rules of physics programmed in so it would sway on curves and jolt on bumpy roads—another task the animators could leave to the software while they concentrated on the characters' acting.

Lasseter brought the look of the film to a new height of realism by deciding that, for the first time in a Pixar production, the rendering process would employ ray tracing. With other rendering techniques—the ones normally used at Pixar and elsewhere—reflective surfaces did not show reflections of nearby objects. Also, standard techniques did not deliver realistic shadows in complex scenes, and light did not refract correctly through materials such as glass or water. The technical directors could cheat a bit to get an image that audiences would find acceptable; in *Toy Story,* for example, on the few occasions when Buzz's helmet showed a reflection, the reflection was a special effect, essentially pasted on.

Ray tracing, on the other hand, simulated the effects of light precisely. It involved following the paths of individual light rays between the light source and the camera. As the virtual rays collided with surfaces—the paint on a car, say—the renderer would track the ray as it bounced off the surfaces or passed through them. The result was as close to reality as a computer could deliver. Although the technique had been known for more than a quarter century, it was rarely used in film production because it soaked up an enormous amount of computer time, even by the standards of computer graphics. (A joke delivered at a SIGGRAPH panel in 1990: "How many ray tracers does it take to screw in a lightbulb?" "Only one, but it takes him eight hours.") Thanks to ray tracing, the frames of *Cars* took an average of seventeen hours apiece to render (and some as long as a week)—to yield one twenty-fourth of a second of screen time—but the reward was to be an unprecedented feel of visual authenticity in a computer-animated feature.

After the long haul ended, the studio rented an art-deco theater in Oakland for the *Cars* wrap party on Saturday, March 11, 2006. The crew indulged through the night and into the next morning—and then they waited. The film's release was still months away.

While the production of *Cars* was winding down, another film, one with an unusually prolonged and complex evolution, remained a frenzy of activity. That film, *Ratatouille,* had its start in early 2000 when Pixar director Jan Pinkava thought of its premise. "I well remember standing in my kitchen with my wife, discussing ideas for stories," he said. "The basic idea of 'a rat wants to become a chef' came to me apparently out of the blue and immediately felt appealing."

Pinkava, born in Prague in 1963, emigrated with his family to England in 1969 following the Soviet invasion of Czechoslovakia.

He joined Pixar in 1993 to direct commercials after having worked for several years for a London-based computer animation house. One of his commercials, made for Listerine, landed Pixar its first Gold Clio, a prestigious advertising award. Later, as director of the 1997 film *Geri's Game,* he brought Pixar its second Academy Award for best animated short film. In the following years, he contributed animation and story work to *A Bug's Life* and *Toy Story 2.*

His concept for *Ratatouille* was a bold, if risky, stroke. Rats in film had mostly been employed for their creepiness—as in the early-1970s films *Willard* and *Ben* and 1989's *Indiana Jones and the Last Crusade.* Walt Disney's own *Lady and the Tramp* presented an intruding rat as a menace that one of the film's canine co-stars had to kill to protect a baby. To conceive a rat as a hero with the run of a kitchen was a considerable leap.

Pinkava finished the first outline of the story by the end of March and collaborated on a script with Jim Capobianco of Pixar's story department. In February 2003, Pixar approved it as the studio's eighth feature—and the first to be released outside its Disney contract. The film had the code name *Project 2006* for its intended release date. Pinkava continued to refine the script together with screenwriters Emily Cook and Kathy Greenberg.

"By the summer of 2004, it had already been a long haul and I needed help," Pinkava said.

In October, with Pinkava's blessing, his colleague Bob Peterson was brought on as co-director. Peterson had established himself as a utility player at Pixar—voice actor, writer, and story artist—and had recently helped Andrew Stanton rewrite *Finding Nemo.*

To Pinkava's dismay, however, the studio gave Peterson exclusive control of the story, while Pinkava remained in charge of other aspects of preproduction, such as character and set design.

In June 2005, Peterson presented his story to a meeting of the studio's "brain trust," as it was known half-sardonically: a group of top directors and story supervisors such as Lasseter, Stanton, Pete Docter, and Joe Ranft who periodically met to comment on works in progress. The presentation went badly; Lasseter regarded the new story as a failure.

Peterson left the project. Pinkava unsuccessfully tried to regain control of the film and would leave Pixar a year later.

Brad Bird had been following the film's progress off and on from the start as a member of the brain trust. In July 2005, Bird was on the second day of a long-awaited vacation when he received phone calls from Jobs, Lasseter, and Catmull asking him to come back and take charge of *Ratatouille.*

The film's release was now planned for the summer of 2007. Bird had eighteen months to revise the story and complete the bulk of production. It was more time than Lasseter had on *Toy Story 2,* but not by much. Bird inherited some useful assets: The digital models for the characters and the most important sets had already been built. Bird retained Pinkava's main characters, including Remy, the ambitious rat; Alfredo Linguini, the inept garbage boy who forms an alliance with Remy; and Anton Ego, the grim, world-weary restaurant critic. He also kept Pinkava's farcical scheme in which Remy would sit on Linguini's head and control his movements by tugging clumps of his hair.

Bird killed the character of Remy's hero chef, Auguste Gusteau, who appears in the film only in television reruns and as a figment of Remy's imagination—acting as a spiritual guide like Obi-Wan Kenobi. Bird retained Remy's brother and father, but simplified the story by eliminating Remy's relationships with his mother and other siblings. "The emotional core of the original story was not to [Bird's] taste," Pinkava said.

> The character of Remy changed profoundly. He became more self-assured and straightforward, confidently following his talent and passion for cooking, and working to overcome the obstacles to his dream. Left behind were other complexities of character: Remy's own struggle with his identity as a rat, his betrayal of his family's values, his awareness of the craziness of his desire to be part of an enemy world, his yearning for acceptance and frustra-

tion with living a lie in the kitchen, all culminating in his final "coming out."

Bird was intrigued by a minor character in Pinkava's story, Colette, a female cook in an otherwise all-male kitchen. He wondered whether such a situation was unusual, and he learned that it wasn't: Serious cooking, he was told, was considered a man's realm in France, and women who wanted a place in it had a high wall to climb. Bird fashioned a much larger role for Colette as a character whose aspirations were being frustrated, and who would ultimately make common cause with Remy and Linguini.

Throughout preproduction, the film's shifting teams carried out the research that had by now become a Pixar signature. They made extensive video footage in the kitchen of an elite Napa Valley restaurant, the French Laundry; Pinkava and producer Brad Lewis served an apprenticeship there. Later, in the interest of accuracy, Bird and a handful of crew members made the sacrifice of spending six days in Paris, taking in its cityscapes and restaurants. To prepare for the sequence early in the film in which the rats travel the fast-moving waters of a storm drain, the effects team rode the rapids of the American River near Sacramento.

In part, Bird's story echoed the themes that had become regular motifs of feature animation since *The Little Mermaid:* Believe in yourself. Celebrate family. As in *The Incredibles,* however, he also took the film's ideas in a more adventuresome direction: Not all individuals are equally blessed with talent. Some people (and some rats), in certain respects, should *not* believe in themselves.

Above all, the script for *Ratatouille* held that where creative talent did exist, whatever its source, it should be treated as precious. Without it comes decline: the slow but sure decline, for instance, of a restaurant that merely recycles the ideas of its past and capitalizes on its founder's fame.

The scenario was an apt metaphor for the Walt Disney Studios that Lasseter and Bird had experienced after CalArts, and for the company that they believed Michael Eisner had presided over in the latter years of his reign. It so happened that Eisner's successor would soon have to decide what price he was willing to pay, and what risks he was willing to take, to bring Lasseter and his team back into Disney's kitchen.

The notion of the Walt Disney Co. acquiring Pixar had come up repeatedly over the years, only to be batted down as soon as it had arisen. Jeffrey Katzenberg had considered and rejected it in the mid-1980s, when Pixar was still an arm of Lucasfilm. In early 1997, in the period between *Toy Story* and *A Bug's Life,* Katzenberg's successor, Joe Roth, tried to sell Eisner on the idea. Roth further urged Eisner to make Jobs—who was not yet back at Apple—his immediate subordinate as Disney's president and chief operating officer. Eisner was decidedly uninterested. In October 2003, a negative cover story on Pixar's stock in *Barron's* pronounced an acquisition unlikely "because the price—$5 billion or more—might strain Disney's balance sheet and dilute earnings."

Iger began considering the idea in a serious way after attending the opening of Hong Kong Disneyland in September 2005. As he watched a parade pass by, he recalled, "I realized there wasn't a character in the parade that had come from a Disney animated film in the last ten years except for Pixar."

The experience brought home to Iger that the Walt Disney Co. was failing badly in feature animation—which Iger believed was Disney's most important business, the one that in the past had created its most enduring films, the one that had supplied the characters for its theme-park attractions and merchandise, the one that had provided many of the songs for its music division, and, above all, the one that had made Disney Disney.

Iger had another concern. Disney's market research showed a worrisome trend: Mothers with children under twelve now rated Pixar's brand higher, on average, than Disney's.

A few weeks after the parade, at a regular Disney board meeting on October 2, Iger brought up for the first time the possibility of acquiring Pixar. With the board's blessing, he contacted Jobs shortly afterward. Over October and November, the two hashed out the pros and cons of such a move; they brought in Ed Catmull, John Lasseter, and the other senior executives of both companies to consider how the combined animation studios might be managed if an acquisition were to take place, focusing especially on how to preserve Pixar's creative culture. Only after those conversations did the two sides begin talking price.

Late in the day on Tuesday, January 24, 2006, four months after his epiphany in Hong Kong, Iger unveiled his response to it: Disney announced that it had agreed to acquire Pixar for 287.5 million shares of Disney stock, then worth roughly $7.4 billion. In theory, the acquisition required shareholder approval, but since Jobs owned 49.8 percent of the shares, his vote was the one that mattered. (The Los Angeles–based investment company TCW Group, the next largest shareholder, held just under 15 percent.)

Disney and Pixar animation would stay in their separate studios in Burbank and Emeryville, but Catmull would become president of the combined organizations. Lasseter would become chief creative officer of both and would serve as principal creative adviser to Walt Disney Imagineering, the unit that Walt had formed in 1952 for the design and engineering of Disneyland, and which now worked on the company's theme parks and other properties around the world. The latter role was apparently in deference to the wishes of Lasseter, the former Jungle Cruise pilot, who advocated tying new theme-park rides to newly released films.

The day after the announcement, Catmull and Lasseter flew down to Burbank and headed for soundstage 7 on the Disney lot. Within the cavernous structure, a crowd of five hundred or so from the ranks of the Disney animation staff awaited an introduction to their future bosses. Applause built as they made their way to the front, and then erupted again in force when Iger and Cook presented them. Lasseter was welcomed as a rescuer of the studio from which he had been fired some twenty-two years before.

Doubtless the audience members hoped that Lasseter's eye would bring commercial success—and, thus, an end to the repetitive layoffs of the past several years. But the applause was plainly about much more than economics. As Ralph Guggenheim had noticed in the early 1990s while staffing *Toy Story,* creative workers valued being part of *important* projects. In feature animation, this meant prestigious films, films respected by one's peers, films popular with one's friends and family, films enduring in memory, films that would someday be looked upon as milestones. Lasseter, it seemed, could be counted on to make Disney animated films important again.

Dave Pruiksma, a co-author of the Disney employees' letter several years earlier in support of Roy Disney, posted a new letter online voicing enthusiasm—to put it mildly—for the new regime.

> The walls of the "Tower Of Terror" have clearly crumbled and "The Ogre King" no longer seems a threat to Walt's Kingdom! Tyrannical heads roll in the hallways as the people again regain control of their kingdom. New, more benevolent leaders have been ushered into power to assure a strong and healthy future and it looks as though the sun shines, once again, upon the happy little kingdom.

Although the acquisition was months away, Catmull and Lasseter quickly began putting their stamp on the operation. They brought back a handful of Disney animation standouts who had only recently been laid off, including Ron Clements and John Musker, co-directors of *Hercules, Aladdin,* and *The Little Mermaid,* and Eric Goldberg, co-director of *Pocahontas.* They shut down the Circle 7 studio and its *Toy Story 3* effort, while bringing most of its 168 artists and staff into other Disney animation projects.

The acquisition formally closed on May 5 after a vote of Pixar's shareholders. Since Pixar had more than $1 billion cash on hand, the net price was $6.3 billion. The figure was steep, if not exorbitant, in relation to Pixar's earnings. For that matter, it amounted to more than 10 percent of the market capitalization of the entire Disney empire: its resorts, its broadcast and cable television networks such as ABC and ESPN, its consumer products licensing, its film library, everything. Moreover, given the rights that Disney already controlled in Pixar's existing films and characters, what it had bought for its $6.3 billion mainly consisted of the services of John Lasseter and the rest of the talent in the studio—and Lasseter was the only one bound by an employment agreement. By traditional measures, Jobs seemed to have taken Iger, and Disney's shareholders, to the cleaners.

Yet traditional measures of value obscured the real importance of Pixar to the Walt Disney Co. The concept of "synergy" was routinely bandied about to justify a merger, but in this case, it was more than rhetoric. Animation was not just the heart of the Walt Disney Co., but its blood and its circulatory system besides. Several weeks after the closing, in an interview with Eisner on Eisner's new CNBC program, Iger made note of that dependence while reflecting on why he had carried out the acquisition.

EISNER: You've made the deal, which I was never able to get done, with Steve Jobs on Pixar. And you made it because you felt animation was the key to the future. I assume that's why you made it, 'cause it was expensive, right?

IGER: Yeah. . . . As you know, animation creates incredible value for the company. You experienced it at the highest level with your—*Little Mermaid*—beginning with *Little Mermaid* . . . all the way through—certainly *Lion King* and movies beyond that. *Lion King* was probably the real high in many respects. And you know that the value that can be created from those films is enormous. And lasts not just our lifetime, but for lifetimes.

We last year reissued the DVD of *Cinderella* which came out in 1950, so a 55-year-old movie. And it sold, I think, over ten million DVD's worldwide.

And I felt—one of the things that I learned from you and [Capital Cities/ABC chairman and CEO] Tom Murphy and others is to know the edge of your own competency. And I felt that we had talent in animation. But we also needed great

> leadership. I didn't think I could provide that leadership in animation. And I believed strongly that the people at Pixar could. Plus they also had tremendous talent.
>
> And so, while it was, you know, fully valued—maybe that's a euphemism—I thought long term for the shareholders of the Walt Disney Company, it was the right thing to do.

Iger would soon have the chance to see audiences react to a new Pixar film. *Cars* opened on June 9, a little more than a month after the transaction. Both commercially and critically, its results were mixed—though only in comparison with Pixar's own past work. Its opening weekend gross, at sixty million dollars, was Pixar's lowest since *Toy Story 2* and brought about speculation that *Cars* would be Pixar's first box-office disappointment. In a sense, the speculation proved correct: *Cars* would actually bring in lower receipts overall, both domestically and worldwide, than any Pixar films since *A Bug's Life* and *Toy Story.* Lasseter's subject matter doubtless played a part in those results, with car racing appealing more to boys than girls, and NASCAR racing in particular less familiar to audiences outside the United States than Formula 1. The disappointment, in any case, was purely relative; although lagging by Pixar standards, *Cars* was still the second-highest-grossing film of the year, after Disney's *Pirates of the Caribbean: Dead Man's Chest.*

Critically, *Cars* was the first Pixar feature to receive a significant stream of negative reactions from reviewers and online commenters. The energy of its race scenes and the beauty and elegant composition of its desert scenery were undeniable, but for many, its script lacked the sparkle of the studio's other efforts. (When Lightning McQueen first sees Sally, his love interest, his exclamation is "Holy Porsche!") At the Web site RottenTomatoes.com, which tracked film reviews, it was the first Pixar film to win less than a 90 percent approval rate (the figure for *Cars* was 76 percent). The reviewer for the *San Jose Mercury News,* the hometown paper of Silicon Valley, lamented, "This is the first Pixar movie with which I've ever found myself getting bored." *The Philadelphia Inquirer* announced, "Pixar finally rolled out a clunker." Animation critic Michael Barrier, a fan of *The Incredibles,* held that *Cars* was "easily the worst of Pixar's features," citing its "clichés, non sequiturs, and stereotypes, ethnic and otherwise." Even with its mixed coverage, however, *Cars* still more than held its own in critical terms against the other computer-animated features the same year.

As it happened, the number of those films, and the number of production companies moving into feature animation, had proliferated wildly. In the decade since the release of *Toy Story,* there had never been more than four major computer-animated releases a year in the United States, most of which came from Pixar, DreamWorks Animation, and Blue Sky Studios. Suddenly, in 2006, the market was so thick with computer animation, mostly aimed at children, that the films seemed to need air-traffic control; among them, in addition to *Cars,* were Animal Logic's *Happy Feet,* Blue Sky Studios' *Ice Age: The Meltdown,* DNA Productions' *The Ant Bully,* DreamWorks Animation's *Over the Hedge,* DreamWorks and Aardman's *Flushed Away,* Nickelodeon's *Barnyard,* Sony Pictures Imageworks' *Monster House* and *Open Season,* and the Disney-distributed film from C.O.R.E. Feature Animation, *The Wild.* Slates for the following years were shaping up to be as long or longer, with still more production houses taking part in the profusion of films.

The trend was easy to bemoan as commercialism at work—executives across Hollywood simultaneously sensing the aroma of money in the air. To be sure, it was that. ("The risk is low, and the upside is amazing," the president of one new studio told *Fast Company* magazine.) Yet it was also further vindication of the vision that Ed Catmull, Alvy Ray Smith, John Lasseter, and others at Pixar had had two decades earlier: 3-D computer-animated films, once far-fetched, now represented the mainstream of family entertainment. (Of course, traditional cel animation in feature films could yet enjoy a resurgence and coexist with the computer variety.)

The increase in computer-animated films also marked the dawning of a democratic moment in artistic expression and entrepreneurship. Just as technological developments in digital production were opening the door more widely in live-action filmmaking, technology was making computer animation more accessible every year.

Computer animation was still an art form that required talent and intense commitment; it wasn't within reach of Everyman. The accessibility of its tools, however, brought new possibilities. Where Pixar's early years had required a succession of wealthy patrons—Alexander Schure, George Lucas, and Steve Jobs—an enterprising artist of the early twenty-first century was not so dependent. The hardware and software of an animator's workstation, once the province of major studios and effects houses, could now be had for the cost of a good used car. As Pixar started its new life as a crown jewel of the Walt Disney Co., it was plausible that it would sooner or later have to jockey release dates with a new kind of rival. Or, rather, it would have to face a rival that looked much the way Pixar itself did thirty years earlier, as a group of men and women in a garage pursuing a dream.

EPILOGUE

After Pixar sold its Pixar Image Computer line in 1990, the acquiring company never built any of the systems. The power of the Pixar Image Computer has since been eclipsed by that of ordinary personal computers, and there are no known examples of the machine still in use. The Computer History Museum of Mountain View, California, has a Pixar Image Computer I in its collection.

Pixar continues to sell PhotoRealistic RenderMan, now known simply as RenderMan. Though it accounts for less than 5 percent of Pixar's revenue, its influence has been out of proportion to its earnings. The digital special effects created by houses such as Industrial Light & Magic and Weta Digital are themselves a form of computer animation, rendered with RenderMan or one of its handful of competitors. So pervasive have digital effects become that ILM—which started out creating motion-controlled models for the original *Star Wars* films—has stopped working with physical models altogether, for lack of demand. Lucasfilm announced in June 2006 that its production unit for effects based on models was to be sold to an employee.

Ratatouille released on June 29, 2007, to mostly rapt reviews. Joe

Morgenstern of *The Wall Street Journal* wrote that it "sustains a level of joyous invention that hasn't been seen in family entertainment since *The Incredibles.*" A minority, such as animator-blogger Michael Sporn, felt so queasy at the very concept of rats running around a kitchen that they found they simply couldn't enjoy the film, however much they admired the animation and writing.

That fall, the company released a retrospective in the form of a coffee-table book through Chronicle Books, titled *To Infinity and Beyond!* The text, written in-house, offered a public-relations version of the company's history—naturally enough. Major characters and events disappeared, while Steve Jobs was alleged to have embraced the vision of Pixar as a feature-film studio from the day he bought the company. Along the way, the book took swipes at Jeffrey Katzenberg, head of DreamWorks Animation, seemingly holding him alone responsible for the story problems in the development of *Toy Story* when he was at Disney. The studio's treatment of Katzenberg in the book was perhaps ungracious considering his role in giving Pixar its entree into feature animation.

Joe Ranft, head of Pixar's story deparment since *Toy Story,* and a voice actor in each of Pixar's first seven films, died on the afternoon of August 16, 2005, during the production of *Cars.* The car in which he was a passenger swerved off a tight curve on the Shoreline Highway and fell over a cliff into the Pacific Ocean. Ranft was enroute to a retreat of the Mosaic Multicultural Foundation; he was a longtime volunteer mentor who reached out to troubled youths at the organization's retreats and evening meetings.

"Sometimes these [retreats] can be difficult—so many men with so many issues, particularly the inner city Black and Latino gang youth," said Luis J. Rodriguez, one of Ranft's fellow mentors. "Joe suffered with us all."

At Pixar, the forty-five-year-old father of two was remembered for his generous and encouraging relationships with his colleagues. Lasseter dedicated *Cars* in his honor.

In 2007 *Forbes* magazine named Steve Jobs the highest-paid executive of any of America's five hundred largest companies, based on gains in the value of stock granted to him at Apple. He was on the board of directors of the Walt Disney Co. Yet his former residence in Woodside, where he had once met with Catmull and Smith and mused about buying Lucasfilm's Computer Division, was now in a state of decay under his ownership.

He had wanted to demolish it; after a group of neighborhood residents opposed his plan to do so, he left the house open to the elements. The interior suffered damage from water and mold. Vines crept up the stucco walls and wandered inside.

The memories that haunted its hallways were those of Jobs's darkest times. He had bought the house only months before the humiliation of his firing from Apple; he lived in it through that firing and through the hard, money-hemorrhaging years of Pixar and NeXT. He left it as his fortunes were about to change, as he was sending Microsoft away from Pixar, convinced that he had something he should hold on to.

When a judge ruled against his quest for a demolition permit, Jobs appealed in 2006 and 2007 all the way to the California Supreme Court, but he lost at every stage. He received proposals from property owners offering to cart the house away in sections and restore it elsewhere; he rejected them. One way or another, it seemed, he meant for the house to be destroyed.

After Alvy Ray Smith left Pixar, founded the graphics software company Altamira, and sold it to Microsoft, he remained as a Graphics Fellow at Microsoft for six years. He retired in 2000, having become wealthy from Microsoft stock options, and divides his time between his homes in Seattle and Bainbridge Island. He was inducted into the National Academy of Engineering in February 2006.

As Ed Catmull entered his seventh decade, he took on not one, but two grand responsibilities. In addition to serving as president of the combined Disney and Pixar animation studios, he became a father again as he and his wife, Susan, adopted a baby boy.

John Lasseter was inducted into the American Academy of Arts and Sciences in October 2007. He awaits the installation of his 1901 steam locomotive and tracks on the grounds of his Glen Ellen, California, property. He has long shared a love for trains with the legendary Disney animators Ollie Johnston, from whom he purchased the locomotive, and Ward Kimball, from whose estate he obtained a seventy-year-old train depot. If his past record means anything, it can be assumed a future Pixar production will portray a locomotive discovering life lessons—once Lasseter hits on the right story. Despite the obligations of his leadership role at Disney, it is difficult to imagine that the boy who emerged into the sunlight after watching *The Sword in the Stone* has directed his last film.

APPENDIXES

ACKNOWLEDGMENTS

NOTES

INDEX

APPENDIX 1
PIXAR ACADEMY AWARDS AND NOMINATIONS

Academy Awards and nominations appear in this list under the year in which the honor was given. The quotations are from the award citations.

Where technical awards have been shared among Pixar employees and employees of other companies, the names of the other companies are noted in brackets.

1987

Best Animated Short Film—Nominee
Luxo Jr.

1989

Best Animated Short Film—Winner
Tin Toy

1992

Scientific and Engineering Award
"To RANDY CARTWRIGHT [Disney], DAVID B. COONS [Disney], LEM DAVIS [Disney], THOMAS HAHN, JAMES HOUSTON [Disney], MARK KIMBALL [Disney], DYLAN W. KOHLER [Disney], PETER NYE, MICHAEL SHANTZIS, DAVID F. WOLF [Disney] and the WALT DISNEY FEATURE ANIMATION DEPARTMENT for the design and development of the 'CAPS' production system for feature film animation."

1993

Scientific and Engineering Award
"To LOREN CARPENTER, ROB COOK, ED CATMULL, TOM PORTER, PAT HANRAHAN, TONY APODACA and DARWYN PEACHEY for the development of 'RenderMan' software which produces images used in motion

pictures from 3D computer descriptions of shape and appearance."

1995

Scientific and Engineering Award
"To GARY DEMOS and DAN CAMERON of Information International, DAVID DiFRANCESCO and GARY STARKWEATHER of Pixar, and SCOTT SQUIRES of Industrial Light & Magic for their pioneering work in the field of film input scanning."

1996

Best Music (Original Musical or Comedy Score)—Nominee
Toy Story (score by Randy Newman)
Best Original Song—Nominee
"You've Got a Friend in Me" from *Toy Story* (music and lyrics by Randy Newman)
Best Screenplay Written Directly for the Screen—Nominee
Toy Story (screenplay by Joss Whedon, Andrew Stanton, Joel Cohen, Alec Sokolow; story by John Lasseter, Peter Docter, Andrew Stanton, Joe Ranft)
Special Achievement Award
"To John Lasseter, for his inspired leadership of the Pixar *Toy Story* team, resulting in the first feature-length computer-animated film."
Scientific and Engineering Award
"To ALVY RAY SMITH, ED CATMULL, THOMAS PORTER and TOM DUFF for their pioneering inventions in Digital Image Compositing."

1997

Scientific and Engineering Award
"To WILLIAM REEVES for the original concept and the development of particle systems used to create computer generated visual effects in motion pictures."
Technical Achievement Award
"To BRIAN KNEP [ILM], CRAIG HAYES [Tippett Studio], RICK SAYRE and THOMAS WILLIAMS [ILM] for the creation and development of the Direct Input Device [aka Dinosaur Input Device]."

1998

Best Animated Short Film—Winner
Geri's Game
Scientific and Engineering Award
"To EBEN OSTBY, WILLIAM REEVES, SAMUEL J. LEFFLER and TOM DUFF for the development of the Marionette Three-Dimensional Computer Animation System."
Scientific and Engineering Award
"To RICHARD SHOUP [Xerox PARC], ALVY RAY SMITH and THOMAS PORTER for their pioneering efforts in the development of digital paint systems used in motion picture production."

1999

Best Music (Original Musical or Comedy Score)—Nominee
A Bug's Life (score by Randy Newman)
Technical Achievement Award
"To DAVID DiFRANCESCO,

BALA S. MANIAN [Digital Optics] and THOMAS L. NOGGLE for their pioneering efforts in the development of laser film recording technology."

2000

Best Original Song—Nominee
"When She Loved Me" from *Toy Story 2* (music and lyrics by Randy Newman)

2001

Academy Award of Merit
"To ROB COOK, LOREN CARPENTER and ED CATMULL for their significant advancements to the field of motion picture rendering as exemplified in Pixar's 'Renderman.' Their broad professional influence in the industry continues to inspire and contribute to the advancement of computer-generated imagery for motion pictures."

2002

Best Animated Feature Film—Nominee
Monsters, Inc.
Best Original Score—Nominee
Monsters, Inc. (score by Randy Newman)
Best Original Song—Winner
"If I Didn't Have You" from *Monsters, Inc.* (music and lyrics by Randy Newman)
Best Sound Editing—Nominee
Monsters, Inc. (sound editing by Gary Rydstrom and Michael Silvers)
Best Animated Short Film—Winner
For the Birds

2003

Best Animated Short Film—Nominee
Mike's New Car

2004

Best Animated Feature Film—Winner
Finding Nemo
Best Original Score—Nominee
Finding Nemo (score by Thomas Newman)
Best Sound Editing—Nominee
Finding Nemo (sound editing by Gary Rydstrom and Michael Silvers)
Best Original Screenplay—Nominee
Finding Nemo (screenplay by Andrew Stanton, Bob Peterson, and David Reynolds; story by Andrew Stanton)
Best Animated Short Film—Nominee
Boundin'

2005

Best Animated Feature Film—Winner
The Incredibles
Best Sound Editing—Winner
The Incredibles (sound editing by Michael Silvers and Randy Thom)
Best Sound Mixing—Nominee
The Incredibles (sound mixing by Randy Thom, Gary A. Rizzo, and Doc Kane)

Best Original Screenplay—Nominee
The Incredibles (screenplay by Brad Bird)

2006

Best Animated Short Film—Nominee
One Man Band

Scientific and Engineering Award
"To David Baraff, Michael Kass and Andrew Witkin for their pioneering work in physically-based computer-generated techniques used to simulate realistic cloth in motion pictures."

Technical Achievement Award
"To ED CATMULL, for the original concept, and TONY DeROSE and JOS STAM [Alias] for their scientific and practical implementation of subdivision surfaces as a modeling technique in motion picture production. Subdivision surfaces have become a preferred modeling primitive for many types of motion picture computer graphics."

2007

Best Animated Feature Film—Nominee
Cars

Best Animated Short Film—Nominee
Lifted

Best Original Song—Nominee
"Our Town" from *Cars* (music and lyrics by Randy Newman)

2008

Best Animated Feature Film—Winner
Ratatouille

Best Original Score—Nominee
Ratatouille (score by Michael Giacchino)

Best Sound Editing—Nominee
Ratatouille (sound editing by Randy Thom and Michael Silvers)

Best Sound Mixing—Nominee
Ratatouille (sound mixing by Randy Thom, Michael Semanick, and Doc Kane)

Best Original Screenplay—Nominee
Ratatouille (screenplay by Brad Bird; story by Jan Pinkava, Jim Capobianco, and Brad Bird)

APPENDIX 2
PIXAR FILMOGRAPHY

The list below excludes trailers (which are often original productions at Pixar), commercials, interstitials, special-effects work, and making-of documentaries.

FEATURE FILMS

Toy Story (1995)
A Bug's Life (1998)
Toy Story 2 (1999)
Monsters, Inc. (2001)
Finding Nemo (2003)
The Incredibles (2004)
Cars (2006)
Ratatouille (2007)

SHORT FILMS

The Adventures of André and Wally B. (1984) [as Lucasfilm Computer Division]
Luxo Jr. (1986)
Flags and Waves (1986)
Beach Chair (1986)
Red's Dream (1987)
Tin Toy (1988)
Knick Knack (1989)
Luxo Jr. in "Surprise" and "Light and Heavy" (1991)
Geri's Game (1997)
For the Birds (2000)
Mike's New Car (2002)*
Exploring the Reef (2003)*†
Boundin' (2003)
One Man Band (2005)
Jack-Jack Attack (2005)*
Mr. Incredible and Pals (2005)*
Mater and the Ghostlight (2006)*
Lifted (2007)
Your Friend the Rat (2007)*

* Made for DVD
† Mixed live-action and computer animation

ACKNOWLEDGMENTS

One of the pleasures of writing a book like this one is getting to know lots of exceptionally smart, creative, and accomplished people. I am grateful to the current and former employees of Pixar who shared their recollections with me, as well as others who were participants in one aspect or another of Pixar's story and who kindly gave of their time. Among them are Ronen Barzel, Nancy Beiman, Loren Carpenter, Jessica Donohoe, Bill Fernandez, Ralph Guggenheim, Pat Hanrahan, Karen Jackson, Alan Kay, Pamela Kerwin, Chuck Kolstad, Tom Lokovic, Nathan Myhrvold, Floyd Norman, Rocky Offner, Mark Oftedal, Fred Parke, Bruce Perens, BZ Petroff, Flip Phillips, Jan Pinkava, Luis J. Rodriguez, Alvy Ray Smith, Adam Summers, and Tom Wilhite. Harrison Price was helpful on the background of the California Institute of the Arts. Lee Van Boven, outside counsel to Lucasfilm in its spin-off of Pixar, helped me understand that transaction. (It goes without saying—but I will say it anyway—that any mistakes of fact or interpretation in this book are solely my responsibility.)

I have also benefited from many written sources, published and unpublished. In this regard, I would like to thank Alvy Ray Smith for generously making his office files available to me. I would also like to note the invaluable work of SIGGRAPH, the Association for Computing Machinery's special-interest group on computer graphics and interactive techniques; for a documentary record of the development of computer graphics in general and computer animation in particular, one can do no better than this organization's journals and conference proceedings. I am indebted to the staff of the Securities and Exchange Commission, which acted with professionalism in determining that various Pixar-Disney and Pixar-Microsoft contracts were no longer entitled to confidential treatment and made them available to me in unredacted form. (Other written sources are indicated in the notes and bibliography.)

Thanks to Emily Spencer for covering Pixar Storytelling Day at Screenwriting Expo 2006 in Los Angeles. Thanks to Judith Hagley and Jonathan Lawlor for helpful editorial comments.

Thanks to Vicky Wilson, my editor at Alfred A. Knopf, for taking a chance on a book about business *and* technology *and* filmmaking and for giving me astute guidance. Also indispensable at Knopf were Kathleen Fridella, Roméo Enriquez, Iris Weinstein, Abby Weintraub, and Carmen Johnson. Thanks to my agents, Glen Hartley and Lynn Chu, for their encouragement and counsel.

I am grateful to Pixar and everyone associated with it for many hours of cinematic pleasure, starting with my viewing of an unfinished version of *Tin Toy* at a conference in the late 1980s.

I want to express my appreciation to my sons, Cameron and Cole, fellow Pixar fans, for striking just the right balance between being patient and not being too patient while I was writing.

Last, but foremost, I thank my wife, Susan, a real-life Elastigirl. Her friendship has meant so much.

David A. Price
September 2007

NOTES

This book is based principally on interviews that I conducted from 2005 to 2007 with participants in Pixar's history at all levels of the company. To keep the notes to a reasonable length, I have cited my interviews only on a sparing basis.

All inflation adjustments in the text are based on consumer price index data from the Bureau of Labor Statistics, U.S. Department of Labor. I used the series CPI-U (all urban consumers).

The abbreviation *int.* in the notes means "author interview"; *pers. comm.* means "personal communication."

FRONT MATTER

v "Character animation isn't": SIGGRAPH course notes, 1994, reprinted in Lasseter (2001), p. 47.

v "All progress depends": George Bernard Shaw, *Man and Superman* (1903), Brentano's ed., 1904, p. 238.

1. ANAHEIM

5 Disney's internal market research: Comments of Robert Iger at Bear Stearns & Co. Annual Media Conference, Feb. 27, 2006.

9 "a feat not of intellect, but of will": Joseph Schumpeter, "The Instability of Capitalism," *The Economic Journal* 38 (Sept. 1928): 379–80.

9 "the joy of creating": Joseph Schumpeter, *The Theory of Economic Development* (1912; reprint, New Brunswick, N.J.: Transaction, 1982), p. 93.

2. IN THE GARAGE

10 a system called Sketchpad: Sutherland (1963).

11 money flooded into the department: The Information Processing Techniques Office of ARPA provided ten million dollars in support for computer graphics at Utah between 1968 and 1975. National Research Council (1999), p. 228.

12 two-year missionary experience: Alvy Ray Smith int.

13 "He just liked to program": Alan Kay pers. comm.

13 a mass layoff: Masson (1999), p. 350.

13 He decided to digitize: Fred Parke int.; Catmull (1972).

14 communication with the Walt Disney Co.: Fred Parke int.; Rivlin (1986), p. 78; Rubin (2006), p. 131.

15 projection of Mickey Mouse: Perry (2001), p. 44.

15 Winnie the Pooh and Tigger: Catmull (1974), picture 22.

15 they were just stepping-stones: *Odysseys in Technology* (2005).

16 Ohio State University: Masson (1999), p. 351.

16 Picture/Design Group: Smith (2001), p. 16; Masson (1999), p. 406; Demos (2005), p. 963.

16 "I had no idea": Auzenne (1994), p. 80.

17 part of a promotional reel: Rubin (2006), p. 105.

17 An enterprising salesman: Ibid.; Smith (2001), p. 15.

17 "Who should it be?": Smith (2001), p. 15.

17 Catmull's computer graphics lab: Alvy Ray Smith int.

18 Alvy Ray Smith background: Alvy Ray Smith int.; Smith (2001), pp. 10–11.

19 Palo Alto Research Center profiled: Stewart Brand, "Spacewar: Fanatic Life and Symbolic Death Among the Computer Bums," *Rolling Stone,* Dec. 7, 1972, pp. 50–58.

20 "I was reluctant to go": Smith (2001), p. 11.

20 SuperPaint: Shoup (2001); Smith (2001), pp. 8–9; Rivlin (1986), pp. 72–75.

20 Smith at Xerox PARC: Smith (2001), pp. 11–13.

22 reminded some of Smokey Bear: Alan Kay pers. comm.

22 "Welcome, California!": Smith (2001), p. 16.

22 "word salad": Ibid.

22 "Our vision will speed up time": Ibid.

25 Computer graphics lab routine: Alvy Ray Smith int.; Ralph Guggenheim int.; Smith (2001), pp. 18–19; Blinn (1998), pp. 132–34.

25 VAX price tag: Thomas B. London and John F. Reiser, *A UNIX Operat-*

ing System for the DEC VAX-11/780 Computer (July 7, 1978) (Bell Labs memo).

25 present-day dollars: With the U.S. consumer price index normalized at 100 for 1983, the CPI figures are for 65.2 for 1978 and 201.6 for 2006.

27 New School lectures, movies, SIGGRAPH: Alvy Ray Smith int.; Blinn (1998), p. 134.

27 "Alex would say": *Odysseys in Technology* (2005).

28 Schure hosted a private showing: Alvy Ray Smith int.

28 "I've just wasted": Ibid.

28 animator named Ub Iwerks: Barrier (1999), pp. 50, 166–68. Iwerks returned to Disney in 1940, hat in hand, and made significant technical contributions in a variety of areas. For his work on special effects in Disney's live-action films, he was honored with Academy Awards in 1960 and 1965. He was also nominated for his special-effects work on Alfred Hitchcock's *The Birds* (1963).

29 making annual pilgrimages: Alvy Ray Smith int.

30 Gindy phone call: Ralph Guggenheim int.

33 NYIT-Lucasfilm visits: Alvy Ray Smith int.

33 Ten minutes or so with Catmull: Rubin (2006), p. 139.

3. LUCASFILM

36 "If you guys are here": Alvy Ray Smith int.

37 a programmer named Loren Carpenter: Loren Carpenter int.; Perry (2001), pp. 44–46; Rubin (2006), pp. 154, 173–74.

37 "They could heat": Perry (2001), p. 46.

37 "It was the closest thing": Auzenne (1994), p. 84.

38 That wasn't the end: Bloomenthal (1998), p. 48.

38 "It became pretty clear": Auzenne (1994), p. 76.

38 simulation of the "Genesis device": Smith (1982); Bloomenthal (1998), pp. 48–49; Alvy Ray Smith int.

41 "Great camera shot": Alvy Ray Smith int. For a technical treatment of the use of particle systems in the Genesis simulation, see Reeves (1983), pp. 97–103.

41 computer-animated displays in *Alien:* Masson (1999), p. 413.

41 *Looker:* Rivlin (1986), pp. 231–32.

42 Star field motion blurred: Loren Carpenter int.

42 and Reeves's fires: Reeves (1983), pp. 99–100.

42 "ballzen shieren offen": Rubin (2006), p. 126.

42 John Whitney, Jr., and Gary Demos: Rivlin (1986), pp. 230–32, 239–40; Masson (1999), p. 407; Demos (2005), pp. 965, 969.

43 Edlund of ILM later showed: Alvy Ray Smith int.

44 *Brave Little Toaster* project: Smith (1984), p. 2.

45 *My Breakfast with André:* Ibid., pp. 1–2.

45 "It got shelved": Schlender (2006), p. 145.

45 "I ran into John": Alvy Ray Smith int.

45 "John, John": Schlender (2006), p. 145.

46 "The key man": Thomas (1958), p. 134.

47 Lasseter background: Lasseter (2004), p. 46.

47 CalArts background: Harrison Price int. At the committee meeting with the nude protester, CalArts board member Stanley Gortikov, head of Capitol Records, looked at the man and said, "What the hell have *you* got to be so proud of?"

48 "It was like being in": Burton and Salisbury (2000), p. 7. Contrary to many, many reports, Lasseter and Burton were not classmates.

49 "We wore those prints out": Cohen (1995), p. 68.

49 Character animation program and student life: Nancy Beiman pers. comm.

50 Burton was one of these: Burton and Salisbury (2000), p. 8.

50 "John's got an instinctive feel": Josh Getlin, "Fate of Next 'Snow White' Rests in CalArts' Hands," *Los Angeles Times,* Oct. 21, 1979, p. V1.

50 "I sense these young kids": Ibid.

51 Walt Disney's portrait: Irwin Ross, "Disney Gambles on Tomorrow," *Fortune,* Oct. 4, 1982, p. 63.

51 staking more than $1.4 million: Barrier (1999), p. 229. It is hard to overstate the historical significance of *Snow White*—but contrary to popular belief, *Snow White* was not the first-ever animated feature. At least two others long predated it: Quirino Cristiani's *The Apostle* (1917), made in Argentina, and Lotte Reiniger's *The Adventures of Prince Achmed* (1926), made in Germany.

51 The company's own market research: Tom Wilhite int.

52 "Now that the cancer": Sito (1998), p. 14.

52 "It was like my heart": Lasseter (2004), p. 46.

52 early footage from *Tron:* The production houses that created computer imagery for *Tron* were Triple-I (Information International, Inc.), MAGI (Mathematics Application Group, Inc.), Robert Abel and Associates, and Digital Effects.

52 "I couldn't believe": John Lasseter, remarks before Visual Effects Society, Feb. 16, 2006.

53 "John was an operator": Tom Wilhite int.

53 stop-motion short called *Vincent:* Burton and Salisbury (2000), p. 15; Tom Wilhite int.

53 "Since it's not": Schlender (2006), p. 145.

55 "Interface Designer": Alvy Ray Smith int.

55 *The Adventures of André and Wally B.:* Smith (1984); Alvy Ray Smith int.; John Lasseter, remarks before Visual Effects Society, Feb. 16, 2006.

58 When SIGGRAPH audiences: Flip Phillips int.; Alvy Ray Smith int.

59 when Wally decides: Lasseter (1987), pp. 38, 40.

59 "He couldn't make the leap": Alvy Ray Smith int.

59 "Even today there is no": Thomas (1984), p. 20.

60 "Today's computers can generate": Thomas (1984), pp. 24–25.

4. STEVE JOBS

61 an engineer named John Isaacs: Solnit (2003), pp. 186–87.

62 Computer described: Levinthal and Porter (1984).

62 the public face of the Lucasfilm: *The Adventures of André and Wally B.* was not the only public face of Lucasfilm at the 1984 conference; the graphics group also presented technical papers, as it did every year at SIGGRAPH.

63 Stock had designed it: Ibid., p. 81. From a technical point of view, the Pixar Image Computer was an early example of SIMD architecture (Single Instruction, Multiple Data).

63 agree on a name: Loren Carpenter int.; Alvy Ray Smith int.; Masson (1999), p. 300.

65 "It was one of those": Jobs (1995), p. 18.

65 "Does not give credit": Linzmayer (1999), p. 71.

65 Jobs was in the habit: Ibid., p. 72.

66 "I told him": Alan Kay pers. comm.

67 estimated net worth: "The Forbes Four Hundred," *Forbes,* Oct. 27, 1986, p. 228.

67 fourteen-bedroom residence: Some details concerning the house are from the Hon. Marie S. Weiner's January 27, 2006, opinion in *Uphold Our Heritage v. Town of Woodside* (Superior Court of California, County of San Mateo No. 444270).

68 lamp on his drawing table: Ralph Guggenheim int.; Flip Phillips int.

69 a Disney engineer named Lem Davis: Alvy Ray Smith pers. comm.

69 Davis eventually persuaded Stan Kinsey: Stewart (2005), p. 57.

70 Over lunch that afternoon: Ibid., p. 55.

70 "Why not let me": Ibid.

70 Eisner and Wells had been inclined: Ibid.

70 excited by a briefing: Alvy Ray Smith, untitled notes, Nov. 14, 1984.
70 Kinsey wanted to go further: Stewart (2005), p. 85.
71 "I can't waste": Ibid.
71 GM management at the time: Patrick Hanratty int. GM's pioneering system, developed under Patrick Hanratty, was called DAC-1, for "Design Augmented by Computer."
72 EDS–Philips deal: Rubin (2006), p. 405; Alvy Ray Smith int.
72 GM's board meeting: Doron P. Levin, *Irreconcilable Differences: Ross Perot versus General Motors* (Boston: Little, Brown, 1999), pp. 255–61.
73 Shortly before Christmas: Rubin (2006), p. 414.
73 If he couldn't sell: Ibid.
74 "like a Porsche": Hertzfeld (2005), p. 29.
75 "I couldn't be bothered": Smith (2001), p. 8.
75 "Steve, this is your workbench": Jobs (1995), p. 3.
76 "It gave one the sense": Ibid., p. 4.
76 Jobs came to hold: Moritz (1984), p. 40.
76 he phoned Bill Hewlett: Ibid., p. 64.
76 introduced him to an older friend: Bill Fernandez pers. comm.
77 "blue boxes": Lundell and Haugen (1984); Moritz (1984), pp. 70–77; Freiberger and Swaine (1984), pp. 207–8.
78 "He wanted money": Moritz (1984), p. 75.
78 visited a friend at Reed: Moritz (1984), pp. 86–87.
78 seeking to adopt him: Steve Jobs commencement address, Stanford University, June 12, 2005.
78 "After six months": Ibid.
78 Jobs background after Reed: Moritz (1984), pp. 89–93, 95–101.
79 "Wouldn't it be neat": Carl Helmers, "What is BYTE?" *BYTE,* Sept. 1975, p. 6.
80 "The theme of the club": Wozniak (1984), p. 74.
80 "Nolan was his idol": Ibid., p. 75.
80 Apple I background: Wozniak (1984); Freiberger and Swaine (1984), pp. 211–13; Moritz (1984), pp. 123–27, 136–45.
81 the largest initial public offering: Moritz (1984), p. 278.
81 Man of the Year: Michael Moritz pers. comm.
81 child support: Moritz (1984), pp. 276–77; Steve Wozniak, "Letters," n.d., woz.org/letters/pirates/02.html.
82 within reach of ordinary users: A history of PostScript and desktop publishing is set out in Pamela Pfiffner, *Inside the Publishing Revolution: The Adobe Story* (Berkeley, Calif.: Peachpit Press, 2003).
82 "You talk to some": Robert Levering, Michael Katz, and Milton

Moskowitz, *The Computer Entrepreneurs* (New York: New American Library, 1984), p. 61.

83 "When Apple was incorporated": Bill Fernandez pers. comm.

83 "Our first computers": Steve Wozniak, "How We Failed Apple," *Newsweek,* Feb. 19, 1996, p. 48.

85 "This whole thing": John W. Wilson, "Look What Steve Jobs Found at the Movies," *BusinessWeek,* Feb. 17, 1986, p. 37.

85 "We thought computer graphics": Alvy Ray Smith pers. comm.

5. PIXAR, INC.

87 "Oh, it's pretty easy": Alvy Ray Smith int.

89 "Who's going to buy": Robertson (1986), p. 61.

89 "People are inherently creative": "The Entrepreneur of the Decade," *Inc.,* April 1989, p. 114.

90 "No matter how short it is": Katie Hafner, "To Infinity and Beyond," *Upside,* Oct. 1, 1997, p. 90.

90 His energies were focused instead: Lasseter (1987), pp. 39, 43.

92 "was the big lamp": Blinn (1998), p. 44. *Luxo Jr.* received an Academy Award nomination for best animated short film, an achievement owed in part to Lasseter's talents and in part to a hacksaw blade. Pixar needed to get a print to Los Angeles by December 1 to qualify for consideration; Lasseter, out of town during Thanksgiving break, began to wonder whether anyone had sent it out. In fact, it was locked in a file cabinet whose owner was also out of town. Craig Good, who had started with the group at Lucasfilm as a janitor and security guard before moving up to a programming job, went into the office over the holiday, bought a hacksaw blade from the hardware store across the street, sawed open the file cabinet, and shipped the application and film off to the Academy.

92 parent lamp as "Dad": Lasseter (1987), pp. 42–43.

92 "portrayal of human sensations": Barrier (1999), p. 121.

92 CAPS background: Alvy Ray Smith int.; Ralph Guggenheim int.; Nancy Beiman pers. comm.; Robertson (1994).

94 upward of 400,000 cels: Thomas and Johnston (1981), p. 317.

95 *Little Mermaid* and *Lion King* multiplane-style shots: Robertson (1994), p. 60, quoting Peter Schneider, then-president of Disney Feature Animation.

95 Jobs irked Catmull and Smith: Alvy Ray Smith pers. comm.

95 RenderMan background: Pat Hanrahan int.; Upstill (1990), pp. xviii–xix; Apodaca and Gritz (2000), pp. 507–9.

99 "Rendering is extremely important now": "Pixar Announces New President, Chairman," Business Wire, Dec. 1, 1988.

99 "Photorealistic three-dimensional images will soon": Martin Marshall, "Pixar Ships Developer's Toolkit for DOS and Unix," *InfoWorld,* Dec. 11, 1989, p. 101.

101 Executives at NeXT received: Brian Dumaine, "America's Toughest Bosses," *Fortune,* Oct. 18, 1993, p. 39.

102 "Everyone begged me": Lasseter (2004), p. 47.

102 "This is a Pixar film": Lasseter (1989), p. 234.

103 "John, you did it": Lasseter (2004), p. 48.

106 "There is a realism": Lawrence M. Fisher, "Computer Animation Now Coming of Age," *New York Times,* April 12, 1989, p. D-1.

106 "going to take over": Ibid.

106 "Where am I going with this?": Ellen Wolff, "Lasseter: Kid in Candy Store," *Daily Variety,* Oct. 30, 1996.

107 "We were aware": Burr Snider, "The Toy Story Story," *Wired,* Dec. 1995, p. 146.

108 "A four-minute masterpiece": Bill Pannifer, "Fine Tooning in 3D," *Independent,* Dec. 6, 1991, p. 15.

108 "probably the closest thing": Bob Swain, "Breathing New Life into Animation," *Guardian,* Nov. 14, 1991.

110 its Listerine spot that year: Mary Huhn, "Batteries and Boxer KO the Competition," *Adweek,* Nov. 12, 1990.

111 through the mid-1990s: Pixar announced on July 8, 1996, that it would close the commercials group so that those employees could work on films and interactive products.

116 "Our goal is to make": Lawrence M. Fisher, "Hard Times for Innovator in Graphics," *New York Times,* April 1991, p. D-5.

6. MAKING IT FLY — I

124 Schneider, however, continued to take: Stewart (2005), p. 165.

125 "As the story evolved": *Toy Story* production notes, Oct. 23, 1995, p. 23.

127 Disney exercised its right: Michael Sragow, "The Toy Story Story: Computer Animators Draw on Technology to Make Their Characters Come Alive," *Sun-Sentinel,* Nov. 25, 1995, p. 1-D.

127 "a great structure": Jim Kozak, "Serenity Now!" *In Focus,* Aug.–Sept. 2005.

129 "What I loved about Tom": *Toy Story* production notes, Oct. 23, 1995, p. 24.

130 "Instead of making Buzz": Ibid.

131 went back and gave the news: During the production break on *Toy Story,* the first commercially marketed, fully computer-animated work emerged from a small Chicago-based firm called Big Idea Productions. Initially sold through Christian bookstores, the thirty-minute video *Where's God When I'm S-Scared?* presented Bible-themed stories with characters in the form of talking, singing vegetables.

131 "He had to wind up": Lasseter and Daly (1995), p. 47.

131 "Who said your job": Ibid.

131 The new script had several changes: Ibid., p. 48.

133 *Toy Story* production pipeline: Ronen Barzel int.; Lasseter and Daly (1995), pp. 42–43; *Toy Story* production notes, Oct. 23, 1995, pp. 36–40; Porter and Susman (2000); Pixar Animation Studios prospectus, Nov. 29, 1995, pp. 35–36.

135 "The eyes more than anything": *Toy Story* production notes, Oct. 23, 1995, pp. 29–30.

135 rejected automatic lip-synching: Porter and Susman (2000), p. 28.

137 the final lighting of a shot: Apodaca and Gritz (2000), pp. 338, 383–409.

137 Buzz and the alien squeak toys: Lasseter and Daly (1995), p. 72.

7. MAKING IT FLY—2

139 Efforts to sell Pixar, talks with Microsoft: Pamela Kerwin int.; Alvy Ray Smith int.; Nathan Myhrvold int.

141 several of Pixar's patents: The patents licensed to Microsoft were U.S. patent numbers 4,897,806, 5,025,400, and 5,239,624 and their worldwide counterparts. Pixar-Microsoft Patent License Agreement, June 21, 1995 (on file with the author).

142 The consumer products arm: Ralph Guggenheim int.; Pam Kerwin int.; John Deverell, "To Infinity and Beyond! World Besieging Local Toy Maker," *Toronto Star,* Nov. 23, 1996, p. E1.

145 Catmull and Lasseter were earning: Edwin Catmull employment agreement, Aug. 1, 1991 (on file with the author); John Lasseter employment agreement, Aug. 1, 1991 (on file with the author).

145 allocation of Pixar stock options: Recapitalization Agreement, Schedule A, April 28, 1995 (on file with the author); Pixar Animation Studios prospectus, Nov. 29, 1995, p. 52.

146 Jobs created a profit-sharing plan: Pixar Cash Profit Sharing Plan, Feb. 22, 1993 (on file with the author).

147 the lawyer who was handling: Larry Sonsini int.; "Larry Sonsini '66," *Boalt Hall Transcript,* spring 1995, p. 48; Katrina M. Dewey, "King of the Valley," *San Francisco Daily Journal,* Oct. 11, 1993; William W. Horne, "A Maverick Matures," *The American Lawyer,* Sept. 1996.

148 For Jobs, Sonsini was not just: Ibid.

148 laid out the problem: Pam Kerwin int.

149 Disney's marketing engine: Jeff Jensen, "Disney Unwraps Brand-New 'Toy'; BK, Frito-Lay Join $145M Push," *Advertising Age,* Nov. 6, 1995, p. 1.

150 next to the El Capitan: The *Toy Story* attraction next to the El Capitan was called "Totally Toy Story" and was open from November 22, 1995, to January 1, 1996. Parts of it later resurfaced as attractions in Disneyland in 1996.

151 critics had already posted: Janet Maslin, "There's a New Toy in the House. Uh-Oh," *New York Times,* Nov. 22, 1995, p. C-3; Richard Corliss, "They're Alive!" *Time,* Nov. 27, 1995, p. 96; David Ansen, "Disney's Digital Delight," *Newsweek,* Nov. 27, 1995, p. 89; Owen Gleiberman, "Plastic Fantastic," *Entertainment Weekly,* Nov. 24, 1995, p. 74; Kevin McManus, " 'Toy': Animation Sensation," *Washington Post,* Nov. 24, 1995, p. N-54; Mark Caro, "A Mesmerizing Trip into Toyland," *Chicago Tribune,* Nov. 22, 1995, p. 1.

154 offices of Robertson, Stephens: Pam Kerwin int.

155 "There was such a palpable passion": *Odysseys in Technology* (2005).

155 "We learned considerably": Ralph Guggenheim pers. comm.

8. "IT SEEMED LIKE ALL-OUT WAR"

159 Disney approved the treatment: Letter from Steve Gerse, Walt Disney Feature Animation, to Samuel Fischer, Ziffren, Brittenham, Branca & Fischer, July 7, 1995 (on file with the author).

159 the job of co-director: Ralph Guggenheim int.; BZ Petroff int.

160 the crew dispersed: Ralph Guggenheim pers. comm.; Ronen Barzel pers. comm.

161 "That one was a really big": BZ Petroff int.

162 "We took out mandibles": *A Bug's Life* production notes, Oct. 10, 1998, p. 14.

163 From a business standpoint: Co-Production Agreement [between] Walt Disney Pictures and Television and Pixar, Feb. 24, 1997.

165 triumphant return to Apple Computer: Linzmayer (1999), pp. 230, 237–38; Peter Burrows, "An Insider's Take on Steve Jobs," *BusinessWeek Online,* Jan. 30, 2006.

165 "the front guy": Joseph E. Maglitta, "My Tough Luck: A Year After Being Fired at Apple, Gil Amelio Talks Candidly," *Computerworld,* July 27, 1998.

166 During Easter weekend 1994: Mike Hoover int.; U.S. National Transportation Safety Board Factual Report—Aviation no. SEA94FA096; U.S. National Transportation Safety Board interview with Michael David Hoover, April 3, 1994; Trip Gabriel, "Survivor," *Outside,* Feb. 1996; letter from Mike Hoover to Bell 206 owners, Sept. 12, 1997.

168 Katzenberg was shocked to learn: Stewart (2005), pp. 139, 158–64.

168 "I am angry because": Ibid., pp. 163–64.

169 "Nineteen years with Michael Eisner": Kirk Honeycutt, "Hollywood's 'Dream Team,' " *Hollywood Reporter,* Oct. 13, 1994.

169 "This is not the first time": Sallie Hofmeister, "Hollywood Falls Hard for Animation," *New York Times,* Oct. 17, 1994, p. D-1.

171 "Jeffrey, how could you?": Peter Burrows, "Antz vs. Bugs," *BusinessWeek Online,* Nov. 23, 1998.

172 "What we were hearing": Pam Kerwin int.

173 The technique, known as subdivision surfaces: DeRose, Kass, and Truong (1998); Apodaca and Gritz (2000), pp. 109–11.

173 "The bad guys rarely win": Amy Wallace, "Ouch! That Stings!" *Los Angeles Times,* Sept. 21, 1998, p. F-1.

173 "a schlock version": Lawrence French, "An Interview with John Lasseter," www.fortunecity.com/skyscraper/pointone/581.

174 *Toy Story 2* background: Ralph Guggenheim int.; Floyd Norman int.

175 estimated hundred million dollars in profits: Jeffrey Daniels, "Theme Parks Put Disney in Third-Quarter Coinland," *Hollywood Reporter,* July 27, 1995.

175 Everything else about the *Toy Story* sequel was uncertain: Karen Jackson int.

178 "You could go to *A Bug's Life*": Karen Jackson int.

178 Roth and Peter Schneider viewed: Steve Jobs shareholder letter, June 1998.

178 In addition to the unexpected artistic caliber: Ralph Guggenheim int.

179 Brannon would also be credited: Strictly speaking, *Toy Story 2* had yet a fourth co-director in Colin Brady, an animator on *Toy Story* who joined Brannon for a time as co-director of the direct-to-video version.

180 "I was sitting in Steve Jobs's": "Collaborative Storytelling Panel," Screenwriting Expo, Oct. 21, 2006.

180 "Everybody was totally entertained": Mark Oftedal int.
181 "John has got a real eye": Floyd Norman int.
181 Other changes included: Ralph Guggenheim int.; Karen Jackson int.; Floyd Norman int.
182 "The characters now interact better": *Toy Story 2* production notes, Oct. 29, 1999, p. 24.
182 also went through revisions: Ibid.
184 "Even though *Toy Story 2*": "Collaborative Storytelling Panel," Screenwriting Expo, Oct. 21, 2006.
184 *Toy Story 2* at CalArts: Angela M. Lemire, "Creative Force: CalArts Gets 'Toy' Preview," *Daily News of Los Angeles,* Nov. 20, 1999.
185 Reviewers found the film: Kirk Honeycutt, "Toy Story 2," *Hollywood Reporter,* Nov. 18, 1999; Todd McCarthy, "Toy Story 2," *Daily Variety,* Nov. 18, 1999, p. 2.

9. CRISIS IN MONSTROPOLIS

187 Lori Madrid had been writing: The discussion of the Lori Madrid lawsuit is based on the court file in *Madrid v. Chronicle Books, et al.,* U.S. District Court for the District of Wyoming, case no. 01-cv-185, filed Oct. 24, 2001. Judge Brimmer's decision granting summary judgment is published in *West's Federal Supplement* at 209 F. Supp. 2d 1227 (D. Wyo. 2002).
195 "All of these ideas": 209 F. Supp. 2d at 1243.
195 Pete Docter began work: The discussion of the evolution of the *Monsters, Inc.* story is based on the court files in *Madrid v. Chronicle Books* op. cit., and *Miller v. Pixar Animation Studios, et al.,* U.S. District Court for the Northern District of California, case no. 02-04748, filed Oct. 1, 2002.
198 "He's like a seasoned lineman": *Monsters, Inc.* production notes, Oct. 16, 2001, p. 31.
198 conversation with his mentor Frank Thomas: Lasseter (2004), p. 48.
199 each main character had its own: Ibid., p. 20.
199 refine the rendering of fur: Lokovic and Veach (2000); Robertson (2001), pp. 24, 26.
200 "For anything with complex": Tom Lokovic pers. comm.
200 The Fizt program also controlled: Baraff et al. (2003); Robertson (2001), pp. 24–25.
201 render farm comparison: Robertson (2001), p. 22; Porter and Susman (2000), p. 26.

207 "It seems that a famous illustrator": Jeffrey Cohen, "On One Eyed Monsters," Feb. 1, 2006, jjcohen.blogspot.com/2006/02/on-one-eyed-monsters.html.

10. EMERYVILLE

208 "When my son was five": *Finding Nemo* production notes, April 29, 2003, p. 14.

210 "Boom, you suddenly cared": Andrew Stanton, "Understanding Story," Screenwriting Expo, Oct. 21, 2006.

210 by the actor William H. Macy: Confidential source int.

210 Dory's character was more: Peter N. Chumo II, "Finding Nemo," *Creative Screenwriting,* Jan.–Feb. 2004, p. 58.

210 "The protagonist's battle": The interview was conducted in April 2004 by a member of the staff of the Christian Film & Television Commission for the 2005 book *So You Want to Be in Pictures?* It was omitted from the book for lack of space and instead appeared online at www.mediawisefamily.com/syw/i-stanton.html. Ted Baehr pers. comm.

211 The underwater setting: Desowitz (2003); Cohen (2003).

211 Summers lecture and consultancy: Adam Summers int.

212 Dory swam without wiggling her tail: Cohen (2003), p. 60.

212 climbed inside a dead gray whale: Adam Summers int.

214 highest-grossing animated film: The box-office figures for *Finding Nemo* are $339.7 million domestic and $866.9 worldwide; for 1994's *The Lion King,* they are $312.8 domestic and $768 million worldwide. Based on U.S. consumer price index data, however, the inflation adjustment from 1994 to 2003 is a factor of approximately 1.23—and the adjusted box-office figures for *The Lion King* are $384.7 million and $944.6 million, respectively, considerably higher than any other animated film up to that time.

214 "five successes out of five attempts": Kenneth Turan, "Hook, Line, and Sinker," *Los Angeles Times,* Calendar, May 30, 2003, p. 1.

214 "*Finding Nemo* Is Pixar's 500th": Chris Suellentrop, "The Geniuses Behind Finding Nemo Are the Next Disney. Uh Oh," *Slate,* June 5, 2003.

214 Miyazaki visit: Suzuki (2003).

217 "Brad would hang out": Dave Gardetta, "Mr. Indelible," *Los Angeles,* Feb. 2005, p. 82.

218 "I kept having these movies": "The Incredibles," Comic-Con 2004, July 23, 2004 (panel).

218 Bird was irked to see: Letter from Brad Bird, *Time,* July 12, 1993, p. 5.

218 Warner Bros. offered him: Bird (1998), p. 21.

220 the company announced a multifilm contract: Marc Graser, "Pixar Plucks Bird, Roth," *Daily Variety,* May 5, 2000, p. 1.

220 "The dad is always expected": Sam Chen, "Brad Bird's Super-Insights on *The Incredibles,*" www.animationtrip.com/item.php?id=257.

221 Bird envisioned her: *The Incredibles* production notes, 2004, pp. 16–17.

222 Bird's original pitch focused: Vaz (2004), p. 125.

223 "The hardest thing": Desowitz (2004), p. 31.

224 "the Pixar glaze": *The Incredibles* production notes, 2004, p. 13.

224 Bird decided that in a shot: Dave Gardetta, "Mr. Indelible," *Los Angeles,* Feb. 2005, p. 80.

11. HOMECOMING

227 second project of the Secret Lab: Floyd Norman int.

229 Eisner, pleased with the sequel's: Ralph Guggenheim int.

229 Eisner testified before the Senate: Brooks Boliek, "Eisner: Piracy 'Killer App' for Computer Profiteers," *Hollywood Reporter,* March 1, 2002; Stewart (2005), pp. 382–84.

230 "Do you know what Michael": Stewart (2005), p. 383.

230 he had come to distrust Eisner: Ibid., p. 2.

230 "I'll never make a deal": Ibid., p. 395.

231 "Yesterday we saw": Richard Verrier and Claudia Eller, "A Clash of CEO Egos Gets Blame in Disney–Pixar Split," *Los Angeles Times,* Feb. 2, 2004, p. A-1; Stewart (2005), p. 408.

231 Jobs offered Eisner terms: Laura Holson, "Pixar to Find Its Own Way as Disney Partnership Ends," *New York Times,* Jan. 31, 2004, p. C-1; transcript of Pixar Animation Studios 2003 Q4 earnings conference call, Feb. 4, 2004.

232 Jobs had met with executives: Laura Holson, "Pixar Executives Not Quite Ready to Forsake Others for Disney Alone," *New York Times,* May 12, 2003, p. C-1.

232 "These were the people": Schlender (2006), p. 140.

232 Gore contacted Disney board member: Claudia Eller and Richard Verrier, "Gore Had Cameo in Disney, Pixar Rift," *Los Angeles Times,* Oct. 10, 2003, p. C-1.

232 Bryson stunned Roy: Stewart (2005), pp. 2–3, 465.

234 "No one has a lock": Bruce Orwall, "Can Disney Still Rule Animation After Pixar?," *Wall Street Journal,* Feb. 2, 2004, p. B-1.

235 "When the wicked witch is dead": Stewart (2005), p. 480.

235 Pixar's quarterly earnings conference call: Pixar Animation Studios 2003 Q4 earnings conference call, Feb. 4, 2004.

239 The company's Web site, meanwhile, ignored: Dominic Jones, "Web-Based Campaigns a Wake-Up Call for Corporations," *IR Web Report,* March 12, 2004.

240 highest no-confidence vote ever: Stewart (2005), p. 513.

241 Eisner himself revealed: Sanford C. Bernstein & Co. Strategic Decisions Conference, June 2, 2004 (transcript on file with the author).

242 called Circle 7: Claudia Eller, "Disney Closes Unit Devoted to Pixar Sequels," *Los Angeles Times,* March 21, 2006, p. C-2; Claudia Eller and Richard Verrier, "Disney Plans Life After Pixar with Sequel Unit," *Los Angeles Times,* March 16, 2005, p. C-1.

242 iPod combined mostly off-the-shelf: Erik Sherman, "Inside the Apple Design Triumph," *Electronics Design Chain,* summer 2002, p. 12. During the fiscal year ending September 24, 2005, according to Apple's SEC form 10-K for the year, Apple sold roughly $4.5 billion in iPod devices and another $899 million in iPod music and accessories, or 39 percent of Apple's $13.9 billion in overall net sales.

243 "It's great to be here": John Markoff and Laura Holson, "With New iPod, Apple Aims to Be a Video Star," *New York Times,* Oct. 13, 2005, p. C-2.

243 Industry analysts expressed amazement: Kim Christensen and Terril Yue Jones, "Launch of Video iPod Shines Light on Jobs, Disney Drama," *Los Angeles Times,* Oct. 13, 2005, p. C-1.

245 "Typically, we'd go into a town": *Cars* production notes, May 15, 2006, p. 31.

246 "While we were at Manuel's": Ibid., p. 32.

246 Porsche: Ann Job, "Lights, Camera, Vroom!" *Detroit News,* April 17, 2006, p. 1-A; "Porsche Goes Hollywood in New Disney–Pixar Movie *Cars,*" Business Wire, June 5, 2006.

247 Underneath the cars' skin: Robertson (2006), pp. 11–12.

247 rendering process would employ ray tracing: Christensen et al. (2006). Turner Whitted, then of Bell Labs, first described ray tracing in a 1980 paper. See Turner Whitted, "An Improved Illumination Model for Shaded Display," *Communications of the ACM* 23, no. 6 (June 1980): 343–49.

248 "How many ray tracers": "SIGGRAPH Bowl," *ACM SIGGRAPH Panel Proceedings* (Aug. 1990), p. 5-2.

248 "I well remember standing": Jan Pinkava pers. comm.

252 Roth further urged Eisner: Anne Thompson, "Losing Nemo," *New York Magazine,* Sept. 15, 2003, p. 24.

252 a negative cover story: Andrew Bary, "Coy Story," *Barron's,* Oct. 13, 2003, p. 21.

252 "I realized there wasn't a character": Comments of Bob Iger at Bear Stearns & Co. Annual Media Conference, Feb. 27, 2006 (transcript on file with the author).

252 Disney's market research showed: Ibid.

252 the possibility of acquiring Pixar: Walt Disney Co. form S-4, Feb. 17, 2006.

254 "The walls of the 'Tower' ": Dave Pruiksma, "A Happy Ending Seems Eminent!" Feb. 2006.

255 "You've made the deal": Bob Iger interview, "Conversations with Michael Eisner," CNBC, May 25, 2006.

256 Critically, *Cars* was the first: Steven Rea, "Pixar's 'Cars' Is No Animated Hot Rod," *Philadelphia Inquirer,* June 9, 2006, p. W-3; Bruce Newman, "Vroom! Pixar's 'Cars' Takes the Back Road, But Be Sure to Slow Down and Enjoy the Ride," *San Jose Mercury News,* June 7, 2006.

257 "The risk is low": Alan Deutschman, "Attack of the Baby Pixars," *Fast Company,* Dec. 2005, p. 61.

EPILOGUE

260 "sustains a level of joyous invention": Joe Morgenstern, "Pixar Cooks with Joy, Inventiveness in 'Ratatouille,' " *Wall Street Journal,* June 29, 2007, p. W1.

260 *Forbes* magazine: "Big Paychecks," *Forbes,* May 21, 2007, p. 112.

261 his former residence: *Uphold Our Heritage v. Town of Woodside* (Superior Court of California, County of San Mateo no. 444270, Jan. 27, 2006); *Uphold Our Heritage v. Town of Woodside,* 147 Cal. App. 4th 587, 54 Cal. Rptr. 3d 366 (Court of Appeal of California, First Appellate District, No. A113376, Jan. 10, 2007), petition for review denied, 2007 Cal. LEXIS 4228 (California Supreme Court, No. S150778, April 25, 2007); Andrea Gemmet, "Steve Jobs Wins Fight to Tear Down Woodside House," *Almanac,* Dec. 22, 2004.

BIBLIOGRAPHY

Apodaca, Anthony A., and Larry Gritz. *Advanced RenderMan: Creating CGI for Motion Pictures.* San Francisco: Morgan Kaufman, 2000.

Auzenne, Valliere Richard. *The Visualization Quest: A History of Computer Animation.* Madison, N.J.: Fairleigh Dickinson University Press, 1994.

Baraff, David, et al. "Untangling Cloth." *ACM Transactions on Graphics* 22, no. 3 (July 2003): 862–70.

Barrier, Michael. *Hollywood Cartoons: American Animation in Its Golden Age.* Oxford, U.K.: Oxford University Press, 1999.

Bird, Brad. "Director and After Effects: Storyboarding and Innovations on *The Iron Giant.*" *Animation World* 3, no. 8 (Nov. 1998): 20–22.

Blinn, Jim. "How I Spent My Summer Vacation, 1976." Chap. 12 in *Jim Blinn's Corner: Dirty Pixels.* San Francisco: Morgan Kaufmann, 1998.

———. "SIGGRAPH 1998 Keynote Address." *ACM SIGGRAPH Computer Graphics* 33, no. 1 (Feb. 1999): 43–47.

Bloomenthal, Jules. "Graphics Remembrances." *IEEE Annals of the History of Computing* 20, no. 2 (April–June 1998): 35–51.

Burton, Tim, and Mark Salisbury. *Burton on Burton* (rev. ed). London: Faber & Faber, 2000.

Catmull, Edwin. "A System for Computer Generated Movies." *Proceedings of the ACM Annual Conference* 1 (Aug. 1972): 422–31.

———. "A Subdivision Algorithm for Computer Display of Curved Surfaces." Ph.D. diss., University of Utah, 1974.

———. "The Problems of Computer-Assisted Animation." *Proceedings of the 5th Annual Conference on Computer Graphics and Interactive Techniques (SIGGRAPH)* 12, no. 3 (Aug. 1978): 348–53.

Christensen, Per H., et al. "Ray Tracing for the Movie 'Cars.' " *IEEE Symposium on Interactive Ray Tracing* 1 (Sep. 2006): 1–6.

Cohen, Karl F. "An Interview with John Lasseter." *Animato!* (fall 1995): 22.

———. "Finding the Right CG Water and Fish in *Nemo*." *Animation World* 8, no. 3 (June 2003): 58–63.

"Computer Research & Development at Lucasfilm." *American Cinematographer* (Aug. 1982): 773.

Demos, Gary. "My Personal History in the Early Explorations of Computer Graphics." *Visual Computer* 21, no. 12 (Dec. 2005): 961–78.

DeRose, Tony, Michael Kass, and Tien Truong. "Subdivision Surfaces in Character Animation." *Proceedings of the 25th Annual Conference on Computer Graphics and Interactive Techniques (SIGGRAPH)*, n.v. (July 1998): 85–94.

Desowitz, Bill. "Depth Perception." *Computer Graphics World* (May 2003): 14.

———. "Brad Bird & Pixar Tackle CG Humans Like True Superheroes." *Animation World* 9, no. 8 (Nov. 2004): 29–33.

Freiberger, Paul, and Michael Swaine. *Fire in the Valley: The Making of the Personal Computer.* Berkeley, Calif.: Osborne/McGraw-Hill, 1984.

Ghez, Didier, ed. *Walt's People: Talking Disney with the Artists Who Knew Him.* 5 vols. Philadelphia: Xlibris, 2005–2007.

Hertzfeld, Andy. *Revolution in the Valley: The Insanely Great Story of How the Mac Was Made.* Sebastopol, Calif.: O'Reilly, 2005.

Jobs, Steven. Oral history. Computerworld Honors Program, April 20, 1995.

Lasseter, John. "Principles of Traditional Animation Applied to 3D Computer Animation." *Proceedings of the 14th Annual Conference on Computer Graphics and Interactive Techniques (SIGGRAPH)* 21, no. 4 (Aug. 1987): 35–44.

———. "Tricks to Animating Characters with a Computer." *Computer Graphics* 35, no. 2 (May 2001): 45–47.

———. "A Tribute to Frank Thomas." *Animation World* 9, no. 8 (Nov. 2004): 46–48.

Lasseter, John, et al. "Bloopers, Outtakes, and Horror Stories of SIGGRAPH Films." *ACM SIGGRAPH Computer Graphics* 23, no. 5 (July 1989): 223–39.

Lasseter, John, and Steve Daly. *Toy Story: The Art and Making of the Animated Film.* New York: Hyperion, 1995.

Levinthal, Adam, and Thomas Porter. "Chap—A SIMD Graphics Processor." *Proceedings of the 11th Annual Conference on Computer Graphics and Interactive Techniques (SIGGRAPH)* 18, no. 3 (July 1984): 77–82.

Linzmayer, Owen W. *Apple Confidential: The Real Story of Apple Computer, Inc.* San Francisco: No Starch Press, 1999.

Lokovic, Tom, and Eric Veach. "Deep Shadow Maps." *Proceedings of the 27th*

Annual Conference on Computer Graphics and Interactive Techniques (SIGGRAPH), n.v. (July 2000): 385–92.

Lundell, Alan, and Geneen Marie Haugen. "The Merry Pranksters of Microcomputing." In Steven Ditlea, ed., *Digital Deli.* New York: Workman, 1984, pp. 57–60.

Masson, Terrence. *CG101: A Computer Graphics Industry Reference.* Indianapolis: New Riders Publishing, 1999.

Masters, Kim. *The Keys to the Kingdom: The Rise of Michael Eisner and the Fall of Everybody Else,* 2nd ed. New York: HarperCollins, 2001.

McCracken, Harry. "Luxo Sr.: An Interview with John Lasseter." *Animato!* (winter 1990): 10.

Moritz, Michael. *The Little Kingdom: The Private Story of Apple Computer.* New York: William Morrow, 1984.

National Research Council. *Funding a Revolution: Government Support for Computing Research.* Washington, D.C.: National Academy Press, 1999.

Odysseys in Technology: A Human Story of Computer Animation. Mountain View, Calif.: Computer History Museum, May 16, 2005. Panel; transcript on file with author.

Parke, Frederick I. "Computer Generated Animation of Faces." *Proceedings of the ACM Annual Conference* 1 (Aug. 1972): 451–57.

Perry, Tekla S. "And the Oscar Goes To . . ." *IEEE Spectrum* 38, no. 4 (April 2001): 42–49.

Porter, Tom, and Galyn Susman. "Creating Lifelike Characters in Pixar Movies." *Communications of the ACM* 43, no. 1 (Jan. 2000): 25–29.

Reeves, William T. "Particle Systems—A Technique for Modeling a Class of Fuzzy Objects." *ACM Transactions on Graphics* 2, no. 2 (April 1983): 91–108.

Rivlin, Robert. *The Algorithmic Image: Graphic Visions of the Computer Age.* Redmond, Wash.: Microsoft Press, 1986.

Robertson, Barbara. "Pixar Goes Commercial in a New Market: Selling the $125,000 Image Processor." *Computer Graphics World* (June 1986): 61.

———. "Disney Lets CAPS Out of the Bag." *Computer Graphics World* (July 1994): 58.

———. "Monster Mash." *Computer Graphics World* (Oct. 2001): 20.

———. "Car Talk." *Computer Graphics World* (June 2006): 10.

Rubin, Michael. *Droidmaker: George Lucas and the Digital Revolution.* Gainesville, Fla.: Triad Publishing, 2006.

Schlender, Brent. "Pixar's Magic Man." *Fortune* (May 29, 2006): 139. Interview.

Shoup, Richard. "SuperPaint: An Early Frame Buffer Graphics System." *IEEE Annals of the History of Computing* 23, no. 2 (April–June 2001): 32–37.

Sito, Tom. "Disney's *The Fox and the Hound:* The Coming of the Next Generation." *Animation World* 3, no. 8 (Nov. 1998): 12–15.

Smith, Alvy Ray. "Special Effects for *Star Trek II:* The Genesis Demo." *American Cinematographer* (Oct. 1982): 1038.

———. "The Making of *André & Wally B.*" Unpublished draft, Aug. 14, 1984.

———. "Digital Paint Systems: An Anecdotal and Historical Overview." *IEEE Annals of the History of Computing* 23, no. 2 (April–June 2001): 4–30.

Solnit, Rebecca. *River of Shadows: Eadweard Muybridge and the Technological Wild West.* New York: Viking Penguin, 2003.

Stewart, James B. *DisneyWar.* New York: Simon & Schuster, 2005.

Sutherland, Ivan Edward. "Sketchpad: A Man–Machine Graphical Communication System." Ph.D. diss., Massachusetts Institute of Technology, 1963.

Suzuki, Toshio, dir. *Thank You, Lasseter-san.* Documentary. Koganei, Japan: Studio Ghibli, 2003.

Thomas, Bob. *The Art of Animation.* New York: Simon & Schuster, 1958.

Thomas, Frank. "Can Classic Disney Animation Be Duplicated on the Computer?" *Computer Pictures* (July–Aug. 1984): 20.

Thomas, Frank, and Ollie Johnston. *Disney Animation: The Illusion of Life.* New York: Abbeville Press, 1981.

Upstill, Steve. *The RenderMan Companion: A Programmer's Guide to Realistic Computer Graphics.* Boston: Addison-Wesley, 1990.

Vaz, Mark Cotta. *The Art of The Incredibles.* San Francisco: Chronicle Books, 2004.

Wozniak, Stephen. "Homebrew and How the Apple Came to Be." In Steven Ditlea, ed., *Digital Deli,* pp. 74–76. New York: Workman, 1984.

INDEX

Page numbers in *italics* refer to illustrations

A NOTE ON THE TYPE

The text of this book was set in Garamond No. 3. It is not a true copy of any of the designs of Claude Garamond (ca. 1480–1561), but an adaptation of his types, which set the European standard for two centuries. It probably owes as much to the designs of Jean Jannon, a Protestant printer working in Sedan in the early seventeenth century, who had worked with Garamond's romans earlier, in Paris, but who was denied their use because of Catholic censorship. Jannon's matrices came into the possession of the Imprimerie nationale, where they were thought to be by Garamond himself, and were so described when the Imprimerie revived the type in 1900. This particular version is based on an adaptation by Morris Fuller Benton.

COMPOSED BY

North Market Street Graphics, Lancaster, Pennsylvania

PRINTED AND BOUND BY

Berryville Graphics, Berryville, Virginia

DESIGNED BY

Iris Weinstein